A DOG'S WORLD

A DOG'S WORLD

Imagining the Lives of Dogs in a World without Humans

Jessica Pierce and Marc Bekoff

PRINCETON UNIVERSITY PRESS
PRINCETON AND OXFORD

Published by Princeton University Press
41 William Street, Princeton, New Jersey 08540
6 Oxford Street, Woodstock, Oxfordshire OX20 1TR

press.princeton.edu

Library of Congress Cataloging-in-Publication Data

Names: Pierce, Jessica, 1965– author. | Bekoff, Marc, author.
Title: A dog's world : imagining the lives of dogs in a world without humans / Jessica Pierce and Marc Bekoff.
Description: Princeton : Princeton University Press, [2021] | Includes bibliographical references and index.
Identifiers: LCCN 2021013530 (print) | LCCN 2021013531 (ebook) | ISBN 9780691196183 (hardback) | ISBN 9780691228969 (ebook)
Subjects: LCSH: Dogs—Behavior—Evolution. | Dogs. | BISAC: NATURE / Animals / Mammals | PETS / Dogs / General
Classification: LCC SF433 .P54 2021 (print) | LCC SF433 (ebook) | DDC 636.7—dc23
LC record available at https://lccn.loc.gov/2021013530
LC ebook record available at https://lccn.loc.gov/2021013531

British Library Cataloging-in-Publication Data is available

Editorial: Alison Kalett, Whitney Rauenhorst, and Hallie Schaeffer
Production Editorial: Natalie Baan
Text Design: Pamela L. Schnitter
Jacket Design: Amanda Weiss
Production: Danielle Amatucci
Publicity: Kate Farquhar-Thomson and Matthew Taylor

Jacket image: Golden retriever in the Rhodope Mountains, Bulgaria / Getty Images

This book has been composed in Simoncini Garamond Std

Printed on acid-free paper. ∞

Printed in the United States of America

1 3 5 7 9 10 8 6 4 2

For Christie Henry

CONTENTS

1 Imagining Dogs in a World without Humans 1

2 The State of Dogs 16

3 The Shape of the Future 40

4 Food and Sex 60

5 Family, Friend, and Foe 82

6 The Inner Lives of Posthuman Dogs 106

7 Doomsday Prepping 126

8 Would Dogs Be Better Off without Us? 143

9 The Future of Dogs and Dogs of the Future 156

Acknowledgments 165

Notes 167

Bibliography 185

Photograph Captions 215

Index 217

A DOG'S WORLD

1

IMAGINING DOGS IN A WORLD WITHOUT HUMANS

A common source of amusement at dog parks, on social media, and in dog-related conversations is just how ridiculously un-wild our dogs can seem. Rufus takes off after a squirrel in the park, running full blast and with an expression of great determination. He reaches the tree well after the squirrel has scampered up the trunk to safety. Maya chases a rabbit; the rabbit pivots and runs left while Maya runs straight forward, oblivious to the rabbit's actual path. Bella barks ferociously at a metal statue of an elk. Poppy stalks a paper bag as it gets blown down the sidewalk by the wind. Dickens refuses to go outside to pee because it is raining. Little Knut shivers uncontrollably when the temperature drops below sixty, despite his tartan sweater. Jethro runs home with his tail between his legs when he senses a wild animal near his mountain home. After antics of this sort, we may shake our heads and remind our dogs just how lucky they are that they have us. "Otherwise," we tell them, "you surely wouldn't survive."

But putting jokes aside, is it true that dogs would be doomed without humans to fill their bowls with kibble, provide shelter from frosty nights or too-hot days, and make sure they don't get themselves into serious trouble? Having both spent years living with dogs whose survival skills seem highly questionable, we have thought about this question off and on and have both spoken sternly to our own dogs about how much they need us. But neither of us had considered the "Would dogs survive without humans?" question in any serious way until chancing upon science journalist Alan Weisman's futuristic eco-fantasy book, *The World without Us*. Weisman asks his readers, "Picture a world from which we all suddenly vanish. Tomorrow."[1] Humans have gone extinct, but everything else—and everyone else—remains. What would happen to your house? To the city in which you bustle back and forth from work, to the grocery, to the gym, to the corner diner? To the ecosystems surrounding your city? To the entire planet, once relieved of the intense pressures of human occupation? And what, we both thought, about dogs?

Weisman's book immediately sparked our curiosity about what life might be like for dogs on a humanless planet. The more we thought about it, the more we wondered whether we might have given our own dogs short shrift and the more certain we were that some or even many dogs would survive and perhaps even thrive in a world without humans. We looked at Weisman's thought experiment of "worlds without humans" through dog-colored glasses, and what we imagined was a world bustling with dog activity, as dogs became part of wild landscapes.

Dogs rarely appear in Weisman's futuristic scenario. This may simply be because his attention was focused elsewhere. But it may also be that he doesn't think their future is all that promising. In one of his few comments about domestic dogs, he proposes that in Manhattan, at least, "wild predators would finish off the descendants of pet dogs" (though "a wily population of feral house cats" will persist by feeding on starlings).[2] The take-away message seems to be that dogs could not and would not survive without us. But is the story of posthuman dogs really this simple and this tragic? We don't think so.

DOGS WITHOUT US

As we began thinking about and researching this book, we started tuning in more carefully to "my dog would never survive without me" conversations and we started taking notes. We were surprised by how frequently people muse about their dog's prospects. We asked friends and strangers what they thought would happen not just to their own dog, but also to dogs in general, if humans were to disappear. Despite some tut-tutting about the hopelessly un-wild behavior of their particular dog, many people gave dogs in general a fighting chance. Here are some of the responses:

"Dogs would be totally screwed."

"Dogs would be fine. They don't really need us all that much."

"Border collies and German shepherds would do great, but Chihuahuas don't have a chance in hell."

"Small dogs would do better because they tend to be feistier and more tenacious than big dogs."

"Clearly large dogs would have the advantage because they will be able to protect themselves."

"Dogs would eventually all be medium-sized."

"Dogs would go back to being wolves."

"Dogs would become like the dingoes in Australia."

"Dogs with 'wild' skills would do better than totally pampered pets."

"They'd learn to survive, even if conditions were bad. Look at the dogs living in the Chernobyl dead zone."[3]

The answers were all over the board. But there were some recurring themes. Many people thought that size would be a major determinant of dogs' success in a world without people, although exactly which size would be best was subject to considerable disagreement. People often mentioned prey drive and hunting skills such as stalking and chasing as determinants of potential survival. Past experiences, particularly time spent as a stray, were mentioned as possible benefits. A good number of people also mentioned a dog's personality, with a confident and fearless dog being given better chances of success than a fearful, overly cautious, or anxious dog.

Would scientists and others who study dogs for a living have similarly diverse intuitions about what might happen to dogs in a world without us? For some clues, we can turn to a 2018 article in *Time* magazine called "How Dogs Would Fare Without Us." Science writer Markham Heid took a stab at the hypothetical dogs-without-humans question, speculating what it would be like for a pampered family dog who suddenly had to survive on her own.[4] Although in Heid's estimation cats are self-reliant and skilled enough to survive without people, many dogs appear "ill-equipped to outcompete other large mammals for food and resources." Is it possible, he asks, that after millennia of domestication, "the entire species may have lost its ability to live independently?"[5]

Heid asked several experts to reflect on this question. Their responses offer some initial scientific speculations about whether dogs would survive and preview some of the key themes we'll be exploring in this book. Most of the experts gave dogs a decent shot at posthuman survival, although they disagreed about the details of which types of dog would survive and which traits might be most adaptive.

For starters, Heid interviewed Alan Weisman, our *A World Without Humans* author. While Weisman was pessimistic about the future survival of dogs, he offered a more nuanced consideration of the future of dogs than he provided in his book. "Dogs aren't too good at fending for themselves," Weisman told Heid, "because we've bred the hunting instinct out of most of them." Most of them would probably not survive, Weisman believes, especially if pitted directly against wild animals such as wolves and coyotes. "The wild animal," declared Weisman, "always wins."[6]

In contrast, Mark Derr, author of *How the Dog Became the Dog*, told Heid that after an initial shakedown period dogs would do quite well. In addition to freely breeding with other dogs, they could also interbreed with wolves and coyotes, because "a horny wolf would not turn his back on a receptive dog."[7] Although small dogs might be more susceptible to predators, they would have certain advantages, such as requiring less food to survive and being able to get away from potential competitors and predators by hiding in small spaces. Small dogs can be incredibly scrappy, too. Derr mentioned the "ferocious" rat terrier who "might do really well hunting and feeding off of small game."[8] Early gangs of dogs would forge alliances to procure food, although these alliances might be less cohesive than wolf packs and more like the looser associations formed by coyotes. Because dogs are adept at forming alliances, they might be willing to cooperate with cats, perhaps even working together to run down and ensnare large game. That dogs are opportunistic feeders and have a broad definition of "edible" will also work in their favor. Natural selection would play a significant role and produce, in time, "a houndy pit-bull type, an 'ur-dog' of around 50–70 pounds."[9] Certain dog breeds would be doomed, by nature of their morphology (their physical form). Derr gives the example of bulldogs, who cannot give birth naturally because the puppies' heads are too large for the mother's birth canal—a result of human breeding practices. "Unless bulldogs learn to give each other cesarean sections, I can't see how they'd make it."[10]

Raymond Pierotti, coauthor with Brandy Fogg of *The First Domestication: How Wolves and Humans Coevolved* and interviewed by Heid, speculated that dogs who are outliers in size would struggle. The largest breeds, including mastiffs, Newfoundlands, and Saint Bernards, would "probably die off quickly because their organs are too small for their body mass." Big dogs are also "too lumbersome to be effective hunters," while very small dogs might wind up being somebody's dinner. Dogs with recent wolf ancestry, such as malamutes, huskies, and Akitas, "would probably do best." The males of these wolf-like breeds may have retained some of the paternal caregiving behaviors that are natural to wolves, but which have largely been lost in pet dogs. Breeds such as border collies, Australian cattle dogs, and hounds that have held onto "ancestral hunting abilities" would also have an edge.[11]

Marc Bekoff, co-author of this book, argued that breed might not be what ultimately matters to survival; more important might be an individual dog's intelligence and skill set. "Some dogs are good hunters," he noted, "while some are good foragers, and some are just really crafty and street-savvy."[12] What would future dogs look like? No one knows, he said. It is unlikely that dogs would resemble their canine ancestors, or that they would become more wolf-like, because this would require more selective breeding than would occur. Nor will dogs become wolf-dog or wolf-coyote hybrids, because you "would need repeat breeding between dogs and wolves or dogs and coyotes for a prolonged period of time, and

I don't think you'd get that." What you would see, Bekoff concluded, is "new and more varieties of *Canis familiaris*."[13] Another change might be that dogs produce fewer offspring overall as they shift from having two reproductive cycles a year to having just one, like wolves and coyotes. After a few generations on their own, the social structure of dog societies might come to resemble the hierarchical and close-knit social structure of wolf packs: "I think dogs would live in groups and have higher- and lower-ranking animals."[14] And while dogs would hunt for prey, they might also scavenge and feed off animals killed by other large predators.

One of the most interesting aspects of Markham Heid's essay was the variety of responses and the range of possible factors that may influence future dog survival. Posthuman dogs are going to be on their own in more important ways than just not having kibble and vet care; they will have to navigate complex ecosystems with which they may be relatively unfamiliar and will have to form relationships with other dogs and other animals with whom they might coexist, cooperate, and compete.

Our own intuitions are in line with those of Markham Heid's experts. We think dogs would survive and even thrive in a posthuman world. And here, in a nutshell, is why: dogs are behaviorally flexible, versatile, and opportunistic (a term used by biologists to mean that an organism can tolerate a wide variety of environmental conditions and will quickly take advantage of favorable conditions when they arise). Moreover, there is already good evidence that dogs can live on their own. Indeed, a relatively small percentage of the billion or so dogs currently living on the planet experience life as "pet" dogs. The majority of the world's dogs don't live within human homes or do so only irregularly. They live as independent individuals, perhaps using human waste as a source of food but not otherwise depending upon humans for social companionship, veterinary care, emotional support, or mental stimulation. The idea that dogs need humans for support and care and that there could be no dogs without their attendant people may simply be wrong.

The most interesting question for us isn't about survival, per se, although who will survive and who won't and why is certainly worth consideration. The exciting question is who dogs will become on their own.

AN EVOLUTIONARY THOUGHT EXPERIMENT

This book is a thought experiment about the survival and evolution of dogs in a humanless future. In embarking on a thought experiment about posthuman dogs, we are connecting with a broader line of inquiry called speculative biology in which scientists make predictions about the trajectory of evolution. The general form of such a thought experiment is "What would happen (or would have happened) *if* . . . ?" What, for example, would have happened *if* dinosaurs hadn't mostly been wiped out by a meteor striking the earth 65 million years ago? (Would humans even have evolved?) Our specific experiment is: "What would happen to dogs *if* humans disappeared?"

Imagine: After roughly 20,000 years of domestication the process abruptly stops, and dogs begin to rewild. What would dogs look like without direct human intervention into breeding? How rapidly would maladaptive traits such as foreshortened snouts be wiped out as natural selection replaced human "artificial" selection? What would dogs eat if bagged dog kibble and human garbage dumps were no longer available to them? Would dogs form groups, and would these be anything like wolf packs in size and social organization? How would dogs who have gone wild reshape the ecosystems within which they are living?

Here are some of our starting speculations about a posthuman dogs' world. Each of these will be examined in detail in later chapters.

- As dogs become whoever they are going to become, it is unlikely that they are going to go back to being wolves. The disappearance of humans would not result in a kind of reverse engineering, where the domestication process rewinds and dogs de-evolve back to who they were before the first wolves tentatively reached out to human beings and vice versa. Posthuman dogs are going to become something entirely, or at least largely, new. The ecological niches that dogs inhabit will be vastly different from the niches that their progenitors filled. The main and most consequential difference is that they will not have human food resources, which may have been one of the key ecological drivers of dog evolution.
- Dogs have been bred for certain physical traits, including the shape and position of ears, the length of tails, and growth patterns and coloration of fur, as well as certain behavioral traits, including a general propensity for friendliness and malleability and breed-specific functional skills such as pointing, fetching, herding, and guarding. Selection for these traits has been driven by an interest in the physical appearance of dogs and by the usefulness of the traits in relation to human pursuits. Taken outside the context of human-canine relations, some of these physical and behavioral traits may serve dogs well. Other traits will likely be downright maladaptive.
- Body size will matter, but one size won't necessarily be better than another. Optimal body size will depend on

what food resources are available, where the dogs live, with whom they share space, and other local conditions.

- Dogs may revert to one reproductive cycle a year, rather than two.
- Some dogs will hybridize with wolves or coyotes.
- Maladaptive phenotypes like short snouts will disappear quickly.
- Dogs will need to solve novel problems related to finding food and staying safe; innovation will be a key driver of success.
- The behavior of current free-ranging dogs is reasonably predictive of how posthuman dogs will behave, at least in the beginning.
- Dogs will be able to adapt to a wide variety of ecosystems.

Speculative biology is an exploration of what *might* be, using out-of-the-box thinking and imagination. But it is grounded in evolutionary theory and existing data and adheres, as much as it can, to scientifically realistic scenarios. In making our predictions about posthuman dogs, we have delved into research on the behavior and biology of canids and, more generally, social carnivores. More than anything, though, we have relied on the growing scientific database on the biology and behavior of dogs, especially the millions of free-ranging and feral dogs who are already living on their own around the world.

The scientific understanding of dogs has grown by leaps and bounds over the past five decades. Yet most of what is known about dog behavior comes from controlled studies of captive dogs in laboratories. Without a doubt, these studies are useful and have helped ground our work. But some of the most intriguing insights—particularly for us in relation to our thought experiment—have come from the scattering of people around the globe who are studying the behavior and social ecology of free-ranging dogs.

Conducting research on free-ranging dogs is challenging; the dogs often have large home ranges, come and go, suffer high mortality (typically human-related), and are often busy at dusk, dawn, and in the dark, when it is hard to see what they are doing. The research can be thankless, too, as free-ranging dogs often are written off as being "feral" rabies-infested pests, neither wild animal (interesting biologically) nor companion (interesting because we love them) but some liminal creature existing in the netherworld between wild and domestic. Research often gets criticized because it is "merely" observational and isn't controlled in the same way as laboratory studies. One field worker told us that he has been ridiculed at scientific meetings because all he does his observe free-ranging dogs and his studies are too uncontrolled to be of any value.

Yet research on free-ranging dogs can help us understand who dogs are and how they make their way in life and will sometimes reveal more than studies of dogs who live in captivity. To take one example, male dogs in captivity are rarely observed playing a role in parenting. But we cannot jump to the conclusion that posthuman male dogs won't be good fathers or won't participate in raising their children. They very well might. As Stephen Spotte remarks in his comprehensive review of free-ranging dog behavior, *Societies of Wolves and Free-ranging Dogs*, "The mere absence of a social phenotypic expression in captivity is not evidence of its extinction, which is why free-ranging dogs make such interesting subjects."[15] Where should we be looking for answers to the questions we're posing about reproductive patterns and other types of behavior in dogs? Not in places where large numbers of dogs have been reproductively neutralized, but rather maybe in the "canine developing world." Ironically, we may learn the most about dogs by looking in places where many dogs are *not* living as pets. And indeed, dogs who are "treated best"—pampered, fêted, fed caviar, and put to bed on Posturepedic mattresses—may be least likely to survive in a posthuman world.

Imagining a future for dogs without their human counterparts helps us shine a light on who dogs are on their own terms, distinct from their cultural role as obedient (or not so obedient) pets, workers, therapists, dumpster divers, and strays. Even more, it asks who dogs might become if humans stopped interfering so completely in their breeding and behavior.

TIME FRAMES AND SCALES OF LOSS

We are setting out to explore what life on Earth would be like for dogs if humans were to exit, en masse, from the planetary scene. We're taking as our main working assumption that all humans disappear all at once, leaving the planet pretty much as it is: habitable, but with significant injury. So, "posthuman" really means "all humans are gone." This is obviously a fictitious scenario. It is highly unlikely that all humans would disappear abruptly, unless there were a massive planetary disaster such as a large meteor strike on a scale that would annihilate all forms of life. Climate change will continue to be felt by nonhuman species, whether humans are around or not. Predicting how global climate change might impact various ecosystems in ten, fifty, one hundred, one thousand, or more years into the future is impossibly complex and so we've mainly left this issue in the background.

A key variable for us is the time frame over which we are considering prospects for canine survival. Things will be significantly different for dogs on day one after humans disappear than at the one-year mark, or after one hundred, one thousand, or ten thousand years. The longer the time scale, the more time natural selection will have to act on remaining dogs. For dogs living in the immediate wake of human disappearance, human selection will still exert an influence on body shape and size, coat type, skull shape, other physical features, as well as on various aspects of behavior. Furthermore, during the first years after human disappearance (give or take twelve to fifteen years), dogs will have lived closely with or around humans, and will have been dependent at least to some degree on humans and on anthropogenic environments. The absence of humans might be felt far more acutely by these dogs than by later generations. After ten to fifteen years, and assuming all humans are gone, dogs will be feral and then eventually they will become wild until they speciate or go extinct.

To emphasize the importance of time frames, we distinguish among Transition dogs, First-generation dogs, and Later-generation dogs. *Transition* dogs are alive when humans disappear and have had some level of human contact. After approximately fifteen years, there will be no more Transition dogs. *First-generation* dogs are born to mothers who had contact with humans. After roughly thirty years, there will be no more First-generation dogs. *Later-generation* dogs are truly posthuman.

THE VALUE OF THINKING ABOUT A FUTURE WITHOUT US

Imagining a future for dogs without their human counterparts is an interesting exercise in biology, but the real value of the thought experiment—and what ultimately motivated us to write this book—is that it can help us think more clearly about who dogs are in the present and this, in turn, can clarify the moral contours

Box 1.1: Nomenclature for Post-Human Dogs

Transition dogs: Dogs who are alive when humans disappear and who have had some level of human contact. After approximately fifteen years, there will be no more Transition dogs.

First-generation dogs: Dogs born to mothers who had contact with humans. After roughly thirty years, there will be no more First-generation dogs.

Later-generation dogs: Truly post-human.

of human-canine relationships. We may find that certain stereotypes ("Stray dogs are starving, lonely, and miserable" or "Dogs are our best friends") are mistaken. Even more, it can help each of us who lives in companionship with a dog answer a question that hovers in the back of our minds: What does it mean to give dogs a good life and, especially, how can I give *my own dog* the best possible life, a life of experiential richness, contentment, and joy?

We have both spent countless hours throughout our careers talking with "dog people" (dog lovers, dog guardians, activists working on behalf of dogs) about how to achieve peaceful coexistence, and a recurring theme in these conversations is that what dogs really want is to be dogs and to be allowed to embody their essential Dogness. Allowing dogs to be dogs and engage in natural dog behaviors means that we need to understand what it means to be a dog—a question that is surprisingly complicated and difficult to answer. One way to answer this question is to take humans out of the picture. An obvious objection is "Oh, you can't do that. Dogs are only dogs in the company of people. A dog's purpose is to be our help-mate, our loyal companion." But is this really a dog's purpose? And isn't this assumption part of our difficulty in thinking clearly about who dogs are? Exploring future scenarios in which dogs are decoupled from humans allows us to gain fresh perspective on the values and commitments of the present. Writing a book about dogs in a world without humans can, perhaps counterintuitively, help us answer the question "How can humans give dogs the best possible life in a world *with* humans?"

In the next chapter, we'll set the stage for our "Who will dogs become without us?" thought experiment by trying to understand who they are now and how they became dogs in the first place. We'll explore what scientists know about the origins of modern domestic dogs and will look for possible clues about the degree to which dogs are dependent upon humans and the extent to which direct human manipulation has "created" these mammals. This may offer us useful information as we speculate about who dogs might become when humans disappear.

2

THE STATE OF DOGS

If humans were to disappear, close to a billion dogs would be left on their own. Who are all these dogs and where do they live? What are their natural behaviors and life-history patterns? How much do they rely on humans for their survival? As we'll see, questions about how dogs would do without us elude easy answers because of the incredible diversity of dogs, the myriad ways in which dogs exploit human habitats, and the surprising dearth of knowledge about who dogs are and how they adapt. Before digging into our exploration of what might happen to all these posthuman dogs in chapters 3 through 6, let's set the stage by thinking about who dogs are now, how much they depend on humans for survival, and what strategies they use to make their way in the world. This, then, can help us think through how dogs might fare without us and how they might evolve in response to a new set of evolutionary pressures.

WHERE DO DOGS SIT ON THE TREE OF LIFE?

Let's move our gaze beyond the individual dogs we know, beyond the furry friend curled up next to us on the couch or impatiently trying to convince us to quit reading and turn attention to the far more interesting task of playing Frisbee. Instead, let's spend a few moments thinking about dogs the way a biologist or zoologist might, locating dogs within the vast expanse of evolved life on Earth and considering who dogs are as animals. This exercise may seem elementary, because every first grader knows that a dog is a mammal with four legs, two eyes, a very active nose, and some sort of tail. But it may surprise you to see just how little we know about our best friends.

So, how would a zoologist look at the species of animals called dogs? Well, the fact is that they simply might *not* look at dogs. There is a tendency among scientists to view dogs as outside the sphere of natural taxa. As a case in point, dogs are often excluded from biological classification schemes, and they rarely appear in zoology textbooks. Luke Hunter's *Carnivores of the World*, for instance, makes no mention of the domestic dog. Although dogs are included in José Castelló's authoritative *Canids of the World*, they warrant only one small paragraph, and are excluded from the phylogenetic tree of the Canidae.[1] (A phylogenetic tree, or evolutionary tree, is a diagram that represents evolutionary relationships among different species.) The invisibility of dogs may seem puzzling, but this omission can be explained by the fact that the phylogenetic tree has traditionally represented only "wild" animals

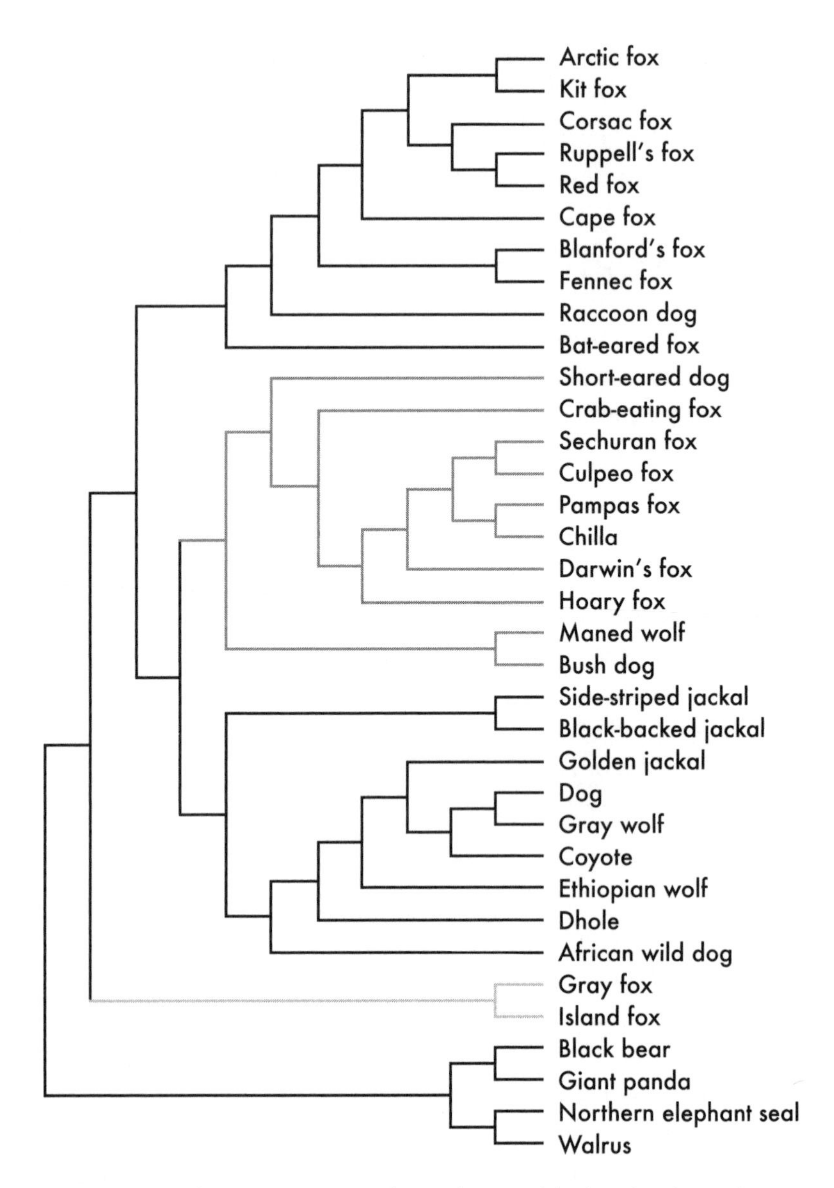

FIGURE 2.1. Phylogenetic Tree of Canidae. Modified and redrawn from K. Lindblad-Toh, C. Wade, T. Mikkelsen, et al., "Genome sequence, comparative analysis and haplotype structure of the domestic dog" (*Nature* 438 [2005]: 803–19).

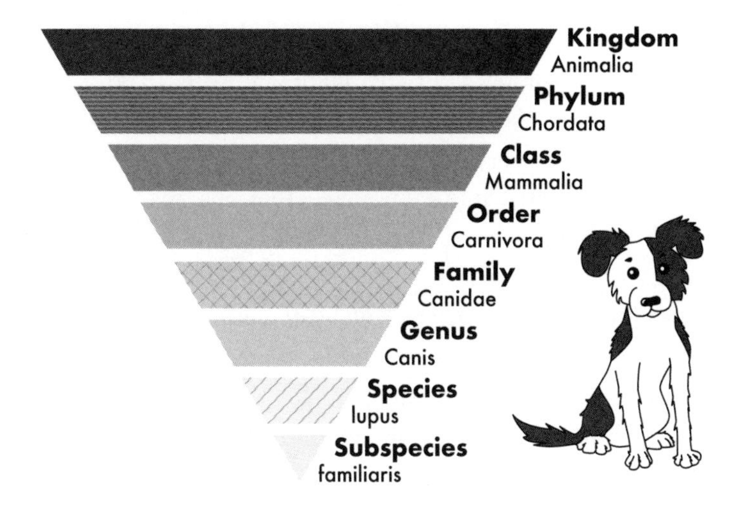

FIGURE 2.2. Taxonomy of Dogs. Within the field of biology, taxonomy refers to the grouping of organisms according to shared characteristics.

that have evolved through natural selection. Accurately or not, dogs have been considered human-made artifacts, products of artificial selection (see figure 2.1).

When dogs do appear on the phylogenetic tree, they are placed within the mammalian order Carnivora ("primarily flesh-eating carnivores"), which is divided into two suborders, the Caniformia (where we find dogs) and Feliformia (where we find cats). Dogs are members of the family Canidae, "a morphologically diverse family of dog-like carnivores."[2] There are thirty-six extant species in the canid family.[3] Dogs are members of the genus *Canis*, along with wolves, coyotes, and jackals (see figure 2.2.).

Scholars disagree about whether dogs should be classified as *Canis familiaris*, indicating that dogs and wolves are separate species, or as *Canis lupus familiaris*, identifying dogs as a subspecies of wolf. *Canis lupus familiaris* is the preferred nomenclature at this point and the convention we follow. There are no subspecies of dog, although there are many different breeds. "Breed" is not a taxonomic classification, but rather an ill-defined term reflecting how humans have chosen to categorize groups of domesticated animals, including dogs. A breed is generally understood to be a group of domesticated animals or plants that have shared behaviors or physical traits that distinguish them from other groups of the same animal or plant.

By looking at the evolutionary soil out of which their family tree sprouted, we've begun gathering some initial clues about how dogs might do in a world without humans. We can gather even more clues by looking at behavioral and physical characteristics of the canids.

WHAT MAKES A CANID A CANID?

What makes a canid a canid as far as biologists are concerned? Canids have distinctive physical characteristics, including their skeletal structure, skull shape, teeth, coats, and limbs. For example, all canids are digitigrade, meaning that they walk on their toes and their heels don't touch the ground when they walk. Their limbs are long and slender and well adapted for running.

Canids also share certain behaviors: living in family groups or small units, cooperation among group members, seasonal breeding, monogamy, parental care, alloparental care (care by aunts, uncles, and unrelated adults), social suppression of reproduction (when the presence of an individual in a group inhibits, behaviorally or physiologically, the reproductive activities of other individuals), dominance hierarchies, long-term incorporation of young adults into social groups, and monoestrous, or one period of "heat" per year. Oxford University's carnivore specialists David Macdonald and Claudio Sillero-Zubiri describe several additional "themes" of canid biology: canids are behaviorally versatile and opportunistic, tend to be highly communicative, and have a propensity to disperse.[4]

Despite shared canid traits, there is wide diversity in how canids look and how and where canids live. One aspect of canid diversity is their size. For example, fennec foxes can weigh two pounds and fit comfortably inside a shoebox, whereas gray wolves can weigh as much as 150 pounds and would have to squeeze into the back of a large SUV. Some canid species have an extremely narrow geographical distribution. For example, Darwin's foxes live only in mainland Chile and on Chiloé Island. Other canids, such as gray wolves and red foxes, can be found on several continents. Canids are found in diverse habitats, from desert ecosystems to arid grasslands, mountainous areas, swamps, large urban cities, and even tundra and ice floes.

Canids also display wide-ranging variations in social behavior that are linked to where and how they live. Wolves typically live in tightly knit packs and hunt cooperatively to bring down the large ungulates that are their main source of food. Although coyotes have been observed living in family groups, they generally live in pairs or on their own and hunt for small mammals alone or together. Foxes tend to live alone and hunt alone but have been seen living in groups.

Canid behavioral patterns can shift depending on ecological circumstances. Coyotes offer an excellent example of how ecological variables can lead to variations in behavior even within the same species. Researchers have found that the availability of winter food strongly influences the social behavior of coyotes in an area during a given time period. In areas where there is enough winter food to go around, coyotes will often form packs that resemble wolf packs.[5] These groups typically are extended families from which some individuals disperse and into which others emigrate. The social structure of these groups might involve occasional spats but is usually maintained by assertive but not necessarily aggressive encounters. When there is not enough food to go around, coyotes typically live in pairs or on their own.

We mentioned that there is considerable diversity in size and shape *among* canids (fennec foxes vs. wolves). Dogs are unique in having extreme *within* species (intraspecific) variation in physical form and behavior. In terms of size, dogs are the most diverse mammal species, with the smallest dogs weighing in at around 4 pounds and the largest weighing up to 175 pounds (or more, if they are slightly chubby from too many treats). The "consistent themes" of canid behavior are found in dogs: they are highly communicative, highly social, versatile, and opportunistic. But there are a few surprises. For example, although one of the themes of canid behavior is social monogamy, dogs can be promiscuous. They are also the only canid in which females typically have two heat cycles a year rather than one.

As canids, dogs have the blueprint for a wide range of behavioral strategies. Local ecological conditions such as prey size, prey availability, and the size of dog populations may shape whether future dogs will behave more like wolves or more like coyotes or foxes or whether they will display novel behaviors and form social systems unlike any of the above.

HOW DOGS BECAME DOGS

Of course, one thing above all else distinguishes dogs from their wild relatives: Dogs are the only canids to have been domesticated. Dogs came from wolves and are genetically extremely close to gray wolves, sharing all but 0.2 percent of mitochondrial DNA.

The origin of modern dogs is still hotly contested among biologists, paleontologists, and anthropologists. Researchers cannot pin down the precise timing of dog domestication, but DNA sequencing and archaeological data place domestication sometime between 40,000 and 15,000 years ago.[6] Dogs were the first animals to be domesticated by a wide margin—roughly 5,000 years—and are likely the only animal domesticated by hunter-gatherers, as other animals were domesticated after the development of agriculture. Some scholars argue that there were probably several domestication events—when humans interfered with natural patterns of breeding—and that wolves evolved into dogs in different places and at different times, whereas others believe there was only one such event.[7] Per Jensen and his colleagues describe dog domestication as "the largest (albeit unconscious) biological and genetic experiment in history."[8]

The scientific questions surrounding how and when dogs evolved from wolves are likely to get muddier before they get clearer. For example, in November of 2019, CNN reported that scientists had discovered the body of a puppy near Yakutsk in eastern Siberia. The dog, named Dogor (which means "friend" in Yakutian), had been preserved by the permafrost and is thought to be about 18,000 years old. Despite extensive DNA tests, scientists cannot decide whether the body belongs to a wolf or a dog or perhaps an animal that was ancestral to both.[9]

Domesticated animals have been selectively bred by humans and are genetically and behaviorally adapted to living alongside us. The key tool of domestication is the control of breeding, and through controlled breeding the selection for traits that are considered desirable. University of Wisconsin philosopher Elliott Sober makes a useful distinction between the selection *for* and the selection *of* traits. Alongside direct selection *for* certain traits—for example, friendliness—there often occurs an indirect selection *of* other unintended traits, or what geneticists would call "hitchhikers."[10] Direct selection *for* hypersociability in dogs may have indirectly introduced other traits, such as changes in pigmentation—

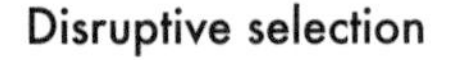

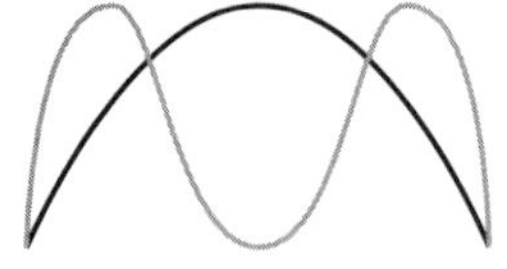

Stabilizing selection

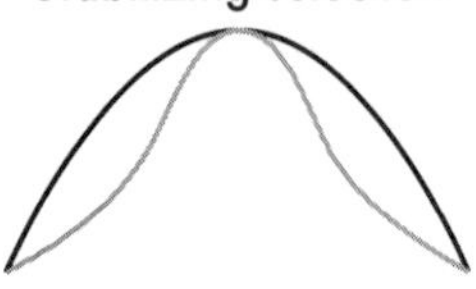

Directional selection

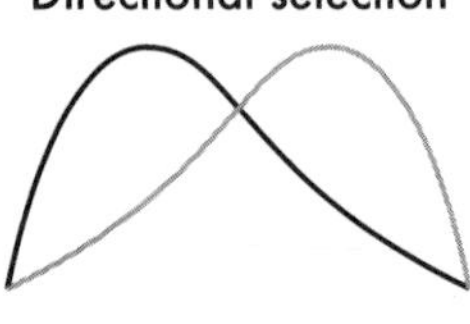

FIGURE 2.3. Three Types of Natural Selection: Simple Definitions for Unfamiliar Readers. Scientists distinguish among three types of natural selection. *Stabilizing selection* is selection for a stable phenotype around a mean (a certain color, running speed, etc.). Dog breeders generally practice artificial stabilizing selection when they try to produce dogs to satisfy breed standards. *Directional selection* is selection for more or less of a given phenotype (e.g., fast running speed, large size, coat color to match ecosystem features). A classic example is the antlers of the now-extinct Irish elk, which grew so large that they couldn't walk because their necks couldn't support the weight. Industrial melanism in moths is another classic example. *Disruptive (or diversifying) selection* is selection at both ends of the normal curve (e.g., for very fast or very slow runners, because those who run at moderate speed get picked off). Redrawn from "Selection Types Chart," by Andrew Z. Colvin (Wikipedia Commons: Attribution-ShareAlike 3.0 Unported CC BY-SA 3.0).

spotted fur or white patches of fur, neither of which is seen in their wild relatives (see fig. 2.3 for more on natural selection).

Anthropologist Darcy Morey describes domestication as an episode of biological evolution.[11] His point is an important, albeit perhaps an obvious one: domestication is a natural process. While it isn't "natural selection" it is nevertheless "natural" in the sense that all biological evolution is natural. Moreover, as we noted above, human control over the evolution of dogs is only partial. We may directly select for certain traits, but evolution doesn't follow simple or straightforward paths because the genetic material upon which evolution works is exceedingly complex. Domestication is still underway: it is a process, not a discrete event.

Domestication has three main impacts on the animals involved, each of which will be relevant to the trajectory of posthuman dogs:

1. Domestic populations are generally much larger numerically than nondomestic populations. Will the presence of great numbers of dogs help them survive?
2. Animals undergoing domestication show distinct physical changes. They are typically smaller than their wild relatives, a side effect of changes in developmental rate and timing due to selection for precocious, or early, sexual maturation. Early sexual maturation is selected for by humans because breeding can occur more rapidly, producing a greater number of offspring and allowing a quicker pace of selection for desired traits. Reduced size is also likely related to the less nutritious diet of domestic animals compared to their wild relatives. Domesticated animals are "paedomorphic," meaning that they retain juvenile features even as adults. Compared to wolves, dogs have shorter snouts, more steeply rising foreheads, smaller sagittal crests, and smaller brains.[12] Which of these physical traits will be adaptive when dogs are on their own, and which might present survival challenges? What will happen to these physical traits under natural selection?
3. Animals undergoing domestication show distinctive behavioral changes, including increased manageability and modifiability of behavior. As Thomas Daniels and Marc Bekoff note, domestication has "enhanced the generalization of bond formation to species other than the domesticant's own (under appropriate conditions) and has led to a lengthening of the socialization period during which social relationships are established."[13] In other words, dogs will bond to humans and there is a longer window of time—the "sensitive period" of socialization from around three to eight weeks of age—during which to nurture this bond. Social inhibitions are muted, making

domesticated animals more neophilic ("likely to approach novel stimuli") than their neophobic ("afraid of the new") wild relatives. Thresholds for avoidance or submissive responses are heightened, meaning that animals are less likely to react with fear or to shy away from new and unexpected situations.[14] How will these behavioral changes affect the social behavior of dogs who have gone wild? Will dogs "repurpose" their sociability and use it to form alliances within groups of dogs or with other animals?

HOW MANY DOGS ARE THERE IN THE WORLD AND WHERE ARE THEY?

The current global population of dogs is about one billion, making dogs one of the most abundant mammals on the planet. Comparing dog population estimates to those for wolves—there are an estimated 300,000 wolves worldwide—shows just how wildly successful dogs have been.[15]

The billion or so dogs on the planet live in a wide variety of ecosystems and have diverse strategies for survival. Dogs can be found on every continent and in nearly every habitable ecosystem—and even in a few that aren't especially habitable.

Where there are dogs, there are also, almost without exception, people. Humans influence the global presence and survival of dogs in direct and indirect ways. People often bring dogs with them to parts of the world where dogs otherwise would not live, such as Antarctica. And people help dogs colonize certain areas in numbers that far exceed the natural constraints of the ecosystem. Some dogs are given a great deal of direct help from humans: they have a particular human or human family from whom they receive targeted affection, food, shelter, and veterinary care. Other dogs don't have primary guardians, but still benefit from the presence of humans by having ample and readily available sources of food, in the form of handouts or scavenged garbage and other human waste. The fact that dogs live where there are humans and rely heavily on anthropogenic (human-generated) resources does not necessarily mean that dogs *must* have humans to survive, just that they usually do.

Population trends for dogs and humans seem to have followed similar trajectories, with both species experiencing explosive growth over the past century. Right now, a conservative estimate would put the ratio of dogs to people at roughly one dog for every ten people.[16]

The geographic distribution of dogs is poorly understood and data are piecemeal.[17] The one billion dogs around the world are not uniformly distributed. Some countries are far more "dog dense" per capita than others. For example, according to data collected by Euromonitor, the U.S. averages one dog for every 4.48 people, while Saudi Arabia has one dog for every 769.23 people.[18] These huge differences in density of dogs may reflect the popularity of pet-keeping practices, cultural attitudes toward dogs, the density of human populations, or some combination of these and other unidentified factors. The U.S. has an estimated 83 million dogs.[19] The unspoken assumption in this figure is that these 83 million dogs are "pets." But the figure of 83 million pet dogs is not adequately descriptive. These 83 million dogs are not all homed or are only homed some of the time and may be free-ranging some of the time. Many are locked in shelters or cycle in and out of homes and on and off the streets. The number of feral dogs living in the U.S. is unknown. Density of dogs varies not only between countries, but also within. Populations of dogs around the world are concentrated in and around urban areas that are densely populated by humans.

In a posthuman world, the geographic distribution of dogs will inevitably shift, and redistributions will occur. Some environments are workable for dogs only because there is support from humans and there will be significant losses for dogs left to fare in these places. Posthuman dogs trying to survive in once-urban environments will face different challenges than those trying to survive in areas that have been less intensively colonized by humans. Their challenges will not necessarily be harder or easier, just different. How densely dog numbers are concentrated, and how much human contact dogs have had, will also shape prospects for survival. Dog-dense areas may provide the advantages of a diverse gene pool and ample opportunities for affiliative interactions, cooperation, and reproduction. In densely populated areas, dogs may experience greater competition for resources and higher levels of interspecific conflict. They may also have to contend with higher levels of transmittable diseases, such as parvovirus, leptospirosis, and rabies.

LIVING ARRANGEMENTS: DOG-HUMAN NICHES

While it's very difficult to know how many dogs there are in the world and where they live, it is perhaps even more challenging to get a handle on the diversity of their living arrangements. Several sources put the number of free-ranging dogs living mainly or entirely on their own at around 80 percent, while about 20 percent of the world's dogs live as "pets," whom we will mainly refer to as homed dogs. To align these percentages with our global population estimates, then, there are about 720 million free-ranging dogs (including stray, street, village, privately owned but free-running, and feral) and about 180 million homed dogs.[20]

When asked, "What is the dog's natural habitat?" many people—scholars and dog guardians alike—answer the question

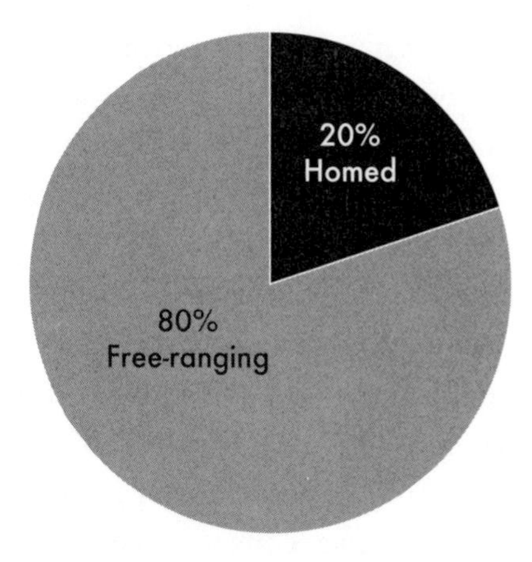

FIGURE 2.4. How Dogs Live.

with "The human home, of course." But as we have just noted, not all dogs live in homes and, in fact, most dogs are not "homed." A more nuanced description of dogs' living arrangement is offered by Darcy Morey, who, in his authoritative book on the domestication of dogs, describes dogs as colonizers of "a new ecological niche, a domestic association with people."[21] Similarly, Per Jensen writes in *The Behavioural Biology of Dogs*, "It has become increasingly obvious that for dogs the most fundamental aspect of the niche it occupies is ourselves. One might therefore say that dogs have come to occupy the ecological niche of living with humans."[22]

Morey and Jensen both point to the critical role of humans in the lives of dogs. And dogs do, of course, occupy niches that overlap significantly with humans, particularly in capitalizing on anthropogenic food supplies. But dogs overlap with humans in many ways, and there are many levels of dependence on humans. Some dogs have high levels of directed or intentional care from humans, relying on people for everything from access to water and food to the opportunity to empty their bladder and bowels. Other dogs live without any direct human help, depending on humans indirectly and only as sources of food. It may be more useful, then, to explore dogs' ecological *niches*.

Are there dogs who are completely independent of us, whose ecological niche does not include or involve humans in some way? In an obvious sense, no dog or any other animal anywhere is totally independent of humans since our activities impact climate and habitats so dramatically and so globally. But there are some dogs—although probably not very many—who are functionally independent, not relying on human garbage or other human-produced food subsidies, and never interacting with humans.

The interesting questions for our thought experiment center on how dogs' current living arrangements and ecological niches will influence future survival. For example, would dogs who are more independent of humans have a better chance at survival if humans disappear? Which ecological niches will dogs occupy in the future when they no longer have the option of the human home or other human-centered habitats?

LABELING DOGS

Researchers use different terminology to categorize dogs and don't necessarily agree with one another on what to call dogs living with or near humans. Terms used to talk about dogs around the world include *homed*, *private*, *poorly supervised*, *uncontained*, *free-ranging*, *streeties*, *community*, *village*, *feral*, *secondarily wild*, and *wild*. We are going to take a quick detour through this terminological landscape. We don't intend to settle any scholarly disputes over which terms are most descriptive ("free-roaming" or "free-ranging"), nor do we need to focus in detail on how many dogs there are in different categories—nobody really knows. For us, the important point is that how dogs are living now will likely make a difference

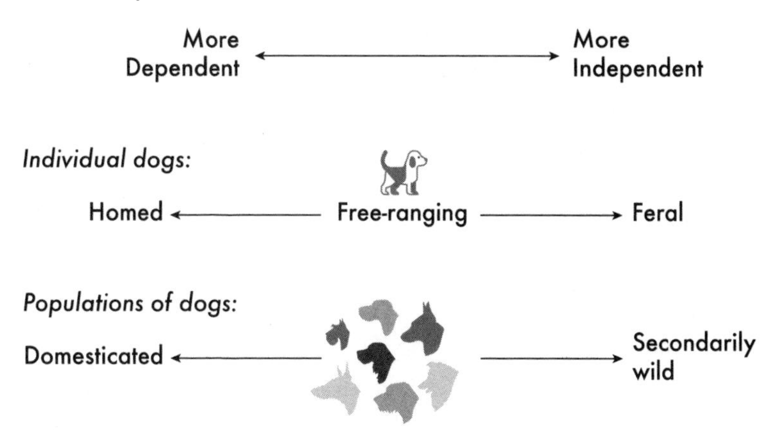

FIGURE 2.5. A Spectrum of Possibilities.

to their future survival without us. For example, one hypothesis is that dogs who have been living a feral life would do better than intensively homed, pampered dogs because they would be "toughened up" and would have honed some useful skills such as hunting and defending food (see figure 2.5).

HOMED

A homed dog is one who lives in a human home. But the designation is quite imprecise. Consider a few different dogs. Hazel is a well-cared-for homed dog. She receives daily food, veterinary care, and sleeps in a warm house on a soft and safe bed, and generally goes outside twice a day when her human family takes her for a short walk so that she can relieve herself. Obie has a human "owner" and home, but he is essentially on his own whenever he wants to be. He goes outside each morning and spends his days traipsing around the neighborhood, returning home at night to get some food and some love (in that order) and to sleep inside on his dog bed. And then there is Sadie, who lives at the end of a chain in her owner's backyard. She is fed irregularly, has no shelter from heat and cold, and her human interaction consists of an occasional friendly pat on the head or unfriendly kick in the side. These would all be labeled homed dogs.

Homed dogs can be more or less free-ranging, and their status as homed might shift over time. They might spend the first five years of life as an indoor-only dog but might then be moved to a home where they have full access to the outside and considerable freedom to roam. Once again, we want to highlight the profound individuality of each dog's experience, and how a dog's circumstances may change dramatically even over their own lifetime.

Classifying a dog as homed may speak to their living arrangement and, importantly, to their psychological and emotional health. For instance, a homed dog who has been recycled through a shelter system four or five times is not the same psychologically as a homed dog who has lived his entire life within the same, stable and loving human household. Psychological variables will play an important role in posthuman survival and adaptation, an issue to which we'll return in chapter 6.

We use the term *pet* colloquially here and there throughout this book, but in terms of classifying dogs, it has little usefulness. *Pet* doesn't have any well-defined meaning. Some pet dogs are loved and treated kindly; some pet dogs are beaten or sexually abused. Some pet dogs have a great deal of freedom and would be classified as free-ranging; others may rarely leave the confines of their home.[23]

FREE-RANGING

The loose category "free-ranging" refers to dogs who have significant opportunities to freely choose where they go and when and includes some homed dogs as well as the many dogs who are otherwise referred to as stray, street, village, poorly supervised, uncontained, or feral. Like homed, free-ranging is an imprecise designation. Some scholars prefer the term free-roaming; we will generally use free-ranging but consider the two terms interchangeable.[24]

Like homed dogs, free-ranging dogs exist across a broad spectrum of independence from humans; some free-ranging dogs interact often and in friendly ways with humans and may even have homes with which they associate. Other free-ranging dogs are well on their way to being feral. For example, science writer Richard Francis describes "village dogs" as individuals who "live outside of human habitations, forage for themselves, and, most important, breed with whomever they choose. . . . Village dogs are hardy creatures that don't owe their survival to human affection."[25] Populations of free-ranging dogs tend to be larger in developing countries and in urban as opposed to rural areas, perhaps because anthropogenic food resources are more plentiful in densely populated human settlements.

FERAL

A feral animal lives in the wild but is descended from a domesticated ancestor. When dogs lose contact with humans, they go through the process of feralization as they adapt to life on their own. The process of feralization refers to changes in *individual* dogs, rather than to changes at the population or species level. Domestication, on the other hand, refers to changes that affect all individuals in a population.[26] Individual dogs do not "de-domesticate" when they're partially or totally out of touch with humans—they feralize.[27]

Feral refers to lifestyle, centering on the near or total absence of human contact. As noted above, the line between free-ranging and feral is fuzzy, and dogs can move back and forth between these rather fluid categories. Often, it's difficult to know if an individual dog has had contact with humans. Puppies born to a feral mother are not necessarily going to be feral themselves. Some might go on to have contact with humans and even might become homed, whereas others might remain feral or pretty much so.

SECONDARILY WILD

The term *secondarily wild* refers to a domestic population that has been free of human-directed selection for long enough that natural selection is acting on all the individuals within the group. How long must a domestic population be under natural selection to become wild? This is an interesting and challenging question. Speciation occurs by degrees. At some point during this slow transition, a border is reached and there will be a difference in kind, and not just degree. How many generations of unfettered reproduction are necessary before a population of dogs is secondarily wild? No one can say for sure. Recalling our earlier discussion of domestication, we can see that the distinction between domestic and wild is fuzzy. There is an ambiguous transition from wild to domestic and an equally ambiguous transition from domestic back to wild.

Very few populations of dogs have thus far become secondarily wild. In *Canids of the World*, José Castelló writes, "There are many commensal and feral populations [of dogs], but the only ones known for sure to be secondarily wild are those on four islands in the Galapagos."[28] Some people consider Australian dingoes another example of a canid who has become secondarily wild. Dingoes have lived and still do live in affiliative relationships with Aboriginal people, but they have always bred freely and without human interference, and many dingoes live completely independently of humans. Like dogs, dingoes evolved from gray wolves and they share many similarities in morphology and behavior with both dogs and wolves. They also have unique traits that are not found in either dogs or wolves, such as flexible joints that allow them to climb trees and rocks and thus better exploit the features of their desert habitat. As fascinating as they are, dingoes are almost certainly not "secondarily wild" because dingoes were probably never fully domesticated. In response to a query about whether dingoes are an example of domesticated dogs gone wild, dingo expert Bradley Smith said that he believes there is more evidence that dingoes were *not* domesticated than that they were. "I think the dingo is a true wild canid with an historical relationship with humans," he said "I don't think they are an example of a dog that's gone feral and reverted to wild again."[29] Based on an analysis of genetic, behavioral, and ecological data, Australian biologist Brad Purcell, who has studied wild dingoes for many years, concluded that "dingoes are not wild dogs . . . dingoes are dingoes."[30]

BREEDS

One of the common labels people use when talking about dogs is "breed" and whether a dog is a purebred or a mutt. Walking down the street with a dog, this is a typical conversation opener: "What a cute dog. What kind is he?" A breed is a group of dogs recognized by a kennel or breed club and which has a pedigree documented in a stud book maintained by these kennel or breed clubs. This is obviously not a biological definition, but one firmly grounded in human culture.[31]

How many of the world's dogs are purebred and how many are mutts? Like our other population numbers, what we have is educated guesswork. Mark Derr estimates that about 30 percent of the world's dogs, or about 300 million, are "pure breed."[32] This is just a global averaging because the proportion of purebred to mutt varies considerably from one country to another. The U.S. is one of the more pedigree-obsessed nations[33] and has a higher-than-average number of purebred dogs, perhaps in the range of 50 to 60 percent.[34]

It seems intuitive to assume that the purebred population would overlap almost completely with the homed dog population and that feral dogs are mostly mutts. But this would be a mistake. Purebred dogs can be free-ranging or feral, and lots of mutts live as homed dogs. Nevertheless, although some feral dogs may be purebred, the odds are extremely high that they would only remain purebred for the first generation.

Breeds are not static constructions or inventions. They are changing continually, even today. The German shepherd of 2020 looks quite different from the German shepherd of one hundred years ago, having a significantly lower croup, wider chest, longer fur, and larger overall size. Breeds may have broad characteristics or temperamental tendencies. For example, border collies have high energy levels and need more physical exercise and higher levels of mental stimulation than are typically provided to dogs living as pets. But no two border collies are alike, and each will respond to a posthuman future in unique ways. Even two border collies who were raised under the same circumstances and exposed to the same environment will respond differently to future events. Although some portrayals of dog breeds, particularly on internet "find your perfect match" sites such as Animal Planet's Dog Breed Selector, seem to suggest that certain breeds of dog have certain personality traits, such as bold, sassy, proud, or brave, these advertisements are misleading. Breeds don't have personalities; individual dogs have personalities.[35]

Some people distinguish between breeds and landraces. A landrace is a group of genetically related dogs unique to a specific geographical location and often serving a function in local human agricultural practices. Landraces are well adapted to local environmental conditions, including altitude, temperature, terrain, and abundance or scarcity of water. Although a landrace shares a certain level of physical uniformity, dogs of a given landrace don't conform to breed standards or belong to breed clubs and registries. One way to think about the difference between landraces and breeds is that landraces have been bred for survival in a particular place, whereas breeds are meant to serve a function for humans, such as pointing at birds, going down rat holes, or having silky fur.

If certain groups of dogs have developed traits that make them adaptive within certain environments, this would seem to give them a leg up on survival when humans disappear, but only if they are living in the ecosystem or one that is closely similar to the one within which they were bred to work and survive. Humans are in the habit of breeding and buying dogs not because they are adapted to the human's home climate or ecosystem, but rather because of some other desired set of behavioral or physical traits. People in Phoenix who buy Bernese mountain dogs don't do so because these dogs are particularly well adapted to living in the desert. Rather, they buy them because they believe they have a mellow disposition or because they like the tricolor coats.

The issue of breed comes into our speculative foray in several ways. First, we might predict that certain breeds would have a survival advantage over others, based on phenotypic traits characteristic of the breed. When we ask people which dogs might survive in a posthuman future, they often answer using specific breeds as an example, and often weaving in size. So, for instance, they'll say that huskies will do better than Chihuahuas because they're larger and will be better hunters and better able to defend themselves against competitors. But, really, there is no reason to think that Chihuahuas will all perish or that huskies will rule the earth.

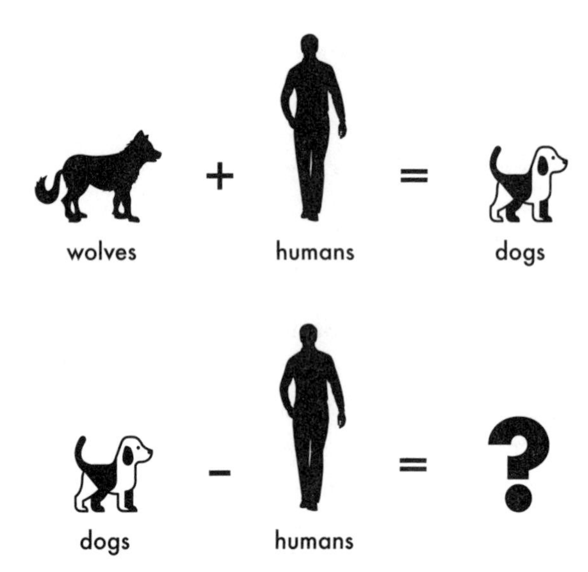

FIGURE 2.6. Looking Backward, Looking Forward. Wolves and humans are the main protagonists in trying to understand how dogs became dogs and, more importantly, what might happen to dogs once we're gone. The human hand is what pushed wolves toward dogs; humans were the secret sauce that, when added to wolves, led to the evolution of dogs. But what happens when dogs' main partner (that's us) is gone? What happens when the process of co-evolution between dogs and humans is abruptly halted by the disappearance of humans, if domestication is stopped in its tracks?

Second, we can ask whether mixed breeds would do better than purebreds, as a rule, in a posthuman world. Again, it is hard to generalize. It will boil down to individual traits and personality.

Third, the inbreeding that currently plagues many purebred dogs may impact future survival because it can lead to genetic mistakes and mutations, and these can manifest as physical malformations such as hip dysplasia and obstructed breathing, as well as psychological problems such as anxiety and obsessive-compulsive disorders.

LOOKING BACKWARD, LOOKING FORWARD

We've outlined who dogs are now in the company of humans and what trajectory has brought them to this point. As we move on to the next four chapters where we explore dogs' feral futures and feral landscapes, one obvious direction to look is backward. For example, is there enough "latent wolf" within dogs such that they will simply go back to where they came from? The answer to this question is easy. Dogs won't go back to being wolves; they will become new and different canids. But who will they become?

3

THE SHAPE OF THE FUTURE

Unless the earth becomes completely uninhabitable, dogs will survive and even thrive in a world without us. But how? Imagining the future of dogs in a posthuman world requires that we cast a very wide net. Among the variables that will influence dog survival are body shape and size, how they get the food they need, how they successfully mate and raise young, how they interact with each other, and how they deal with the ups and downs of life on their own. In the following four chapters, we'll explore these key components of survival for posthuman dogs (see table 3.1: "Ecological and evolutionary trajectories of posthuman dogs").

Table 3.1. Ecological and evolutionary trajectories of posthuman dogs

Functional morphology	Body size
	Temperature regulation/geographical distribution
	Intelligence
	Longevity
	Skull morphology
	Noses
	Ears
	Eyes
	Tails
	Skin
	Coat
	Pigmentation
	Coloration
	Shedder?
	Body type/shape
	Slender or broad
	Hip and pelvis
	Length of extremities
	Allometry
	Gender
Food, feeding ecology, and feeding strategy	Food availability
	Distribution
	Seasonal variation
	Size (size of dog, dimorphism; size of prey species)
	Biomass
	Picky eater?
	Daily caloric needs (typical eating pattern)
	Diet (optimal nutrition/meal plan)
Reproduction	In relation to food supply
	Frequency of cycles
	Years of reproductive fecundity/potential
	Number of offspring
	Mortality of offspring
	Sex ratio
	Mating system (monogamous, polygamous)
	Age at maturity for males
	Age at maturity for females
	Alloparental care
	Denning
	Mortality
Mechanisms influencing population size and dynamics	Dispersal
	Food availability (biomass and its distribution; carrying capacity of ecosystem)
	Clumped or spread-out food?
	Mating patterns
	Robustness of ecosystem (e.g., climate change, potential degradation)
	Presence of competitors (coexist, compete, cooperate)

(continued)

Table 3.1. (continued)

Communication (interspecific and intraspecific)	Visual signals and displays Auditory Olfaction and scent marking Contact behaviors Habitat type
Social organization and social dynamics	In relation to food supply In relation to timing of reproduction Are there family groups? Dispersal patterns Optimal size of group Sex ratio in group Division of labor Group/individual hunting Care of young Defense of territory Dominance hierarchies Leadership Control of intergroup/interspecific aggression
Socialization processes	Parent-infant allegiance and kinship Infant-infant (littermates) allegiance and kinship Sibling (nonlittermates) allegiance and kinship Philopatry (attachment to particular location, or "locality imprinting") Gender
Social relationships	How these change in response to mortality, emigration, birth of offspring Between members of same sex (female-female; male-male) Between members of opposite sex Based on age Based on social rank Parent-young relationships
Use of space	Territorial? Shelter Denning Home range size Ratio of biomass of animal to biomass of available food
Relationships with sympatric species	Prey Predators Commensals Parasites 3 Cs (coexistence, cooperation, competition)

Table 3.1. (continued)

Smarts	Multiple intelligences Street smarts Past experiences Numerical competency
Problem solving	Cooperation and coordination Time and energy budgets Food Shelter Mates Allies Navigation Hunting Frustration Time to "giving up" Ability to self-regulate Resolving conflicts Age Past training (for pet dogs) Exposure to/history with humans
Learning	Learning styles Flexibility and adaptability Social learning Cultural variations Genes and environment/individual predispositions Metacognition Imitation and mimicry/emotional contagion Cooperation and trust Learning unfairness/fairness Resentment, holding grudges Play Teaching by parents, adults, and others
Cognitive ecology	Environmental/ecological challenges and cognition Cognitive ecology and communication
Emotional intelligence	Theory of mind Empathy Emotional contagion Play
Social cognition	Social dynamics Cooperation Coordination Behavioral plasticity Leadership

(continued)

Table 3.1. (continued)

Personality	Difference between personality and temperament Reactive Bold or shy Risk takers or risk averse Persistence Curiosity Predictable vs. unpredictable environments Playfulness Sociability Aggressiveness Curiosity Fearfulness/fearlessness Confidence Openness Facilitator/leader
Coping strategies and response to stress	Different types of stress and eustress Coping styles Reactivity Stability Resilience

We'll be focusing in large measure on what evolutionary biologists call life-history traits and life-history strategies. A life-history strategy, as defined in a classic paper by Stephen Stearns, is "a set of co-adapted traits designed, by natural selection, to solve particular ecological problems."[1] The life history of an organism is how it grows, survives, and reproduces throughout its lifetime. Some examples of life-history traits are litter or brood size, age at reproductive maturity, reproductive lifespan, body size, and diet. When biologists conduct life-history studies their goal is to understand how organisms balance the competing energetic costs of survival, growth, and reproduction—what biologists refer to as evolutionary "trade-offs."

Biologists haven't traditionally done life-history analyses of dogs or other domesticated animals because these animals have been considered artifacts who weren't subjected to natural selection and who thus haven't evolved trade-offs among life-history traits.[2] Nonetheless, we argue that dogs, along with other domesticated animals, *do* have life histories, *have* made necessary trade-offs, and *are* worthy of study. The life-history strategies of dogs reflect the ways in which humans have deliberately manipulated trade-offs among different variables to suit human interests; they also reflect the fact that humans don't actually exert as much control as we think.

Attention to life-history strategies of dogs is a critical piece of our speculative journey. It is through the lens of life-history analysis that we are trying to understand what might happen to dogs if humans were taken completely out of the equation. Through this lens, we can ask, for instance, what might happen when natural selection *alone* is acting on the physical shape and size of dogs. What life-history strategies would converge with those of other canids—and which canids? And how long would it take for these evolutionary changes in form and behavior to take place, once human-directed selection is replaced by natural selection? How would different variables, such as body size and feeding strategy, interrelate? Might dogs succeed because they are starting off with much more behavioral and morphological variation than other species of canid?

Biologists are also interested in what are called phenotypes. Phenotypes are observable traits, both physical and behavioral, that result from interactions between genes and environments. Dogs, like many other mammals, display phenotypic plasticity, which means that a genome can give rise to a range of physiological, morphological, or behavioral traits. Phenotypic plasticity allows organisms to respond and adapt to the social and nonsocial environments in which they live. Phenotypic plasticity also complicates the project of trying to predict how individuals of a given species—in this case *Canis lupus familiaris*—will respond to evolutionary pressures. Dogs around the world live in diverse ecological contexts, and there is an incredible range of individual variation. Some generalizing is hard to avoid, but we will emphasize the huge phenotypic diversity among dogs around the world and will refocus attention on individuals.

WHAT WILL POSTHUMAN DOGS LOOK LIKE?

To discuss what future dogs will look like in the absence of humans, we need to dig into morphology, the branch of biology concerned with the physical form of animals and plants, including their shape, size, and structure. More specifically, studies of functional morphology analyze how particular forms and variations in form are related to survival and reproduction. Some aspects of morphology are evident to the naked eye, such as the variation in size between Great Danes and Pekingese. Other aspects, such as the shape of the brain, the form of different muscles, and the size of the sagittal crest on top of the skull, are less obvious but equally important.

Many, if not all, posthuman dogs will eventually look different from current dogs, as their physical form evolves in response to the pressures of natural selection rather than human selection. Body sizes and shapes that are maladaptive will disappear, and dogs will develop bodies and physical traits such as ears, noses, and coats that are best adapted to the specific demands of climate and feeding strategies, at the very least. That dogs will be different is a given. However, exactly how the physical appearance of dogs might change over time is a surprisingly difficult question to answer, in part because there is enormous diversity in how dogs look and in how physical form influences behavior right now.

Indeed, as we've emphasized, we find more phenotypic variations among dogs than in any other species of animal, wild or domestic.

This unique diversity has resulted from human interference in reproduction, which has led to the expansion of a wide variety of traits well beyond limits normally imposed by the evolutionary pressures of natural selection.

Not only have humans controlled the expression of phenotypic traits of dogs and shaped their evolution for human purposes rather than for survival, but we've also transplanted dogs into a variety of geographic locations independent of how well suited they are morphologically to these environments. Whether their physical form matches the demands of the ecosystems in which dogs find themselves will be largely a matter of luck, especially for Transition and First-generation dogs.

DOES SIZE MATTER?

When we asked people which dogs would survive in a posthuman world, many offered answers related to size. Even people with minimal background in biology jumped to body size as a key factor in survival, and their intuition is right on target: body size is one of the most significant variables in studies of functional morphology and life-history trade-offs.

There are some good arguments on both sides of the "size" coin. Small dogs might do better because they won't need as much food and could burrow and hide from predators and other sorts of danger. Yet small dogs might become prey to a broad range of predators and wouldn't have the size and strength to protect themselves. Large dogs, in contrast, could fight off a wider range of predators and, because of their size, hunt and kill a wider range of potential prey. Being bigger also means that a dog would have more cushion if food became scarce, because larger bodies can store up greater reserves. Yet larger animals tend to have higher caloric needs than smaller individuals, so the benefits of being able to capture larger prey may be offset by the need to eat more. Being bigger would only be advantageous up to a point. Extra-large dogs are plagued by musculoskeletal defects and other maladaptive traits that would make survival on their own challenging.[3]

Broadening out from dogs for the moment, across other carnivorous mammals, variations in a whole range of life-history strategies are often linked to differences in body size. Among canids, for example, body size is linked to reproductive strategy, birth weight, litter size, age of young at weaning, age of independence, age of sexual maturity, and longevity. And as we'll explore in the next two chapters, adult body size is also related to diet and to social organization, with larger body size being linked to higher levels of sociality and group living, both of which are adaptations for subduing large prey.[4]

Body size can also vary in relation to geography and climate, and size differences in dogs might very well influence posthuman survival for Transition and First-generation dogs. Over time, populations of dogs will adapt to specific geographical constraints and subspecies of dogs might eventually evolve.

Ecologists have developed complex algorithms to analyze how variations in climate and geography influence the evolution of species over time, particularly how they influence body size and shape. The application of these ecogeographical "rules" can aid us in imagining how this long-range adaptation might occur and, ultimately, whether big or little or medium-sized dogs will prevail.[5]

Bergmann's rule predicts that body size increases with higher latitude for endothermic, or warm blooded, vertebrates—including humans, dogs, and most other mammals—who maintain constant body temperature regardless of where they live. This rule is thought to hold both across and within species. Higher latitudes are typically colder, so the prediction is that larger-bodied animals will do better than smaller-bodied animals in colder climates, perhaps because larger-bodied animals are generally better at conserving heat than smaller-bodied animals. Over the long term, then, the evolutionary trend for posthuman dogs might be toward larger-bodied subspecies of dog living in Canada and Siberia, and smaller-bodied subspecies living in regions closer to the equator.

Biologists also have a rule for predicting relationships between body shape and climate. Allen's rule predicts that animals living in cold climates will have shorter limbs than animals living in hot climates. Dogs living in cold climates might evolve to resemble current-day huskies or Akitas, with relatively short legs, while dogs living in hotter places might resemble greyhounds or salukis. Of course, the life-history traits of dogs have been heavily manipulated by human selection, so dogs won't necessarily follow ecological rules that might pertain to animals under natural selection, at least in the short term.

Evolution tends to drift toward larger body size over time, at least for mammals. So, in the very long term, dogs might gradually get larger. Unfortunately, larger mammals also tend to go extinct. Moreover, there is good evidence that smaller body sizes can better adapt to heat stress, which is likely to be a considerable advantage on our warming planet. Ecologists have already documented a slow shift toward smaller body sizes in response to climate change, and this pattern may accelerate. A 2019 study by Robert Cooke, Felix Eigenbrod, and Amanda Bates, for example, predicted that the limited set of viable ecological strategies available to mammals and birds will further shrink in the next one hundred years. "Based on species' extinction probabilities," they write, there will be a shift toward "small, fast-lived, highly fecund, insect-eating, generalists."[6] So, maybe dogs will gradually get smaller.

In pretty much any future we can imagine, climatic disruption will be a significant variable in survival. The growing body of research into which life-history traits will help animals adapt to rapid climate change can help present and future ethologists hone their predictions about posthuman dogs. Some mammals can "behaviorally escape" climate change much better than others,[7] and dogs will likely be among the more resilient species, at least in part because of the diversity of body sizes found among dogs—some sizes are bound to "work" even as others may not. Beyond the diversity in body size, dogs have the added advantages of behavioral flexibility, flexibility in activity times (they can be active day or night), and a generalist diet.

VARIATIONS IN BODY SHAPE

Dogs come in as many different shapes as they do sizes. There are short, stout dogs like Staffordshire terriers, pencil-thin dogs like whippets, and sausage-shaped dachshunds. Yet while there are theory-driven predictions about how body size of dogs might influence strategies of adaptation, it is harder to know how body shape might benefit or harm dogs of the future and how it might evolve over time.

In *Domestication: The Decline of Environmental Appreciation*, Helmut Hemmer describes the development of what he calls slender and broad types. These forms are most evident, he says, in horses. Consider, for instance, the difference between a sleek thoroughbred and a heavy-breed horse such as a Clydesdale. Thoroughbreds are built for speed, while draft horses are bred for strength and power. These growth types are also evident in dogs. As in horses, the slender-type dogs, such as greyhounds and whippets, are bred for speed, while broad-type dogs such as the Saint Bernard and bullmastiff are built for strength and power.[8]

The slender and broad types represent functional trade-offs. A trade-off is basically a compromise: an animal gets something that works and is adaptive, but there are other things that the animal "gives up." With respect to dogs, for example, you can't have the speed and agility of a whippet and at the same time have the strength and power of a mastiff. Who will survive better, the fast and agile whippets or solid and strong mastiffs? The answer is "it depends." Small and skinny might be adaptive in some respects (hiding, not needing as much food, greater dexterity), but being broad and stout gives the advantage of muscular strength, and perhaps the upper hand in conflicts that involve fighting. Dogs might develop niche-specific skills, with "power" dogs hunting and eating certain types of prey (larger prey that need to be overpowered) or excelling in certain types of ecosystems (dense undergrowth), whereas "speed" dogs might favor other niches (open landscapes) and hunt for different types of prey (rabbits, mice, and insects).

A dizzying array of skeletal variations and combinations can be observed in modern dogs, all of them involving functional trade-offs such as longer, thinner limbs versus shorter, thicker limbs; large skull on a small body versus small skull on a large body.[9] Yet there are also modern breeds of dog for whom selection to achieve a certain physical form may compromise physical health. For example, the extremely shortened limbs of breeds such as corgis and basset hounds may compromise movement without any apparent functional benefits. Even if short-legged dogs manage to survive the transition, it is hard to imagine that natural selection will retain this body shape.

Canids are cursorial animals (from the Latin *cursor*, or runner): they are born to run. Indeed, canids have been called one of nature's elite endurance athletes.[10] As in humans, there is great variation in dogs' athleticism. Like their human athlete counterparts, some dogs are built like runners, with lithe, long bodies, while some are most assuredly not built like runners. Good runners are not only fit, but also have an economy of movement and gait that makes them able to cover ground efficiently.

In a 2017 study on the effects of domestication on locomotor gait and economy, Caleb Bryce and Terrie Williams hypothesized that more "wolf-like" dogs, including the northern breeds, such as Alaskan malamutes and Norwegian elkhounds, would have greater aerobic economy—they could go longer, harder, stronger—than breeds whose physical bodies are less like their wild canid relatives.[11] Their data confirmed their hypothesis.

SKULLS

The shape of an animal's skull, including the snout, eye sockets, jawbone, and teeth influences how and what they see, smell, hear, and taste, as well as how they acquire and process food. Consider, for example, a very general comparison between carnivore and herbivore skulls. In carnivores, the eye sockets typically face forward to help them home in on prey. Herbivores, who are typically prey animals, have eye sockets on the side of the head, positioning the eyes for greater peripheral vision so they can see who's around them. Carnivores have sharp teeth, including the canines and incisors, for biting and tearing flesh, while herbivores have flattened teeth for chewing plants.

Researchers studying the size and shape of dog skulls are interested in how domestication may have changed the shape of the head, along with dogs' brains, teeth, eyes, and noses. Among the traits that differentiate wolves and dogs is the size of the sagittal crest, a bulge that sits on the top of their skull. The larger sagittal crest of wolves allows for a much more powerful bite. It's possible that as wolves became dogs, the reduction in the size of the sagittal crest and the power of their bite came about because they were being trained to bring prey back to humans rather than consume what they caught.[12] One possibility for posthuman dogs is that their sagittal crest will again get bigger, giving them the needed jaw musculature for being efficient predators. It's also possible that there might be changes in dogs' teeth, so they become better at tearing up food.

Many other changes to the shape of the skull have occurred during the evolution of wolves into dogs. One of the most notable is the appearance in domesticated dogs of what early ethologists called the "infantile schema," or paedomorphism. Dogs have a marked rounding of the head along with a high forehead and large eyes. These paedomorphic traits make dogs appear more baby-like than wolves, even as adults. Scientists hypothesize that paedomorphic traits appeal to humans, make dogs less threatening to us, and stimulate our caregiving impulses. As you can see by looking at a book that features photos of various dog breeds, some have more strongly paedomorphic features than others. A shih tzu is "baby-faced" while a German shepherd has a "wolf-like" face.

The most strongly paedomorphic dog breeds are also brachycephalic, a reference to the short (*brachy*) length of the head (*cephalic*), measured from front to back (nose to neck). These "smushed-face" breeds include French bulldogs, pugs, and boxers. Over time, breeders have been pushing certain dogs toward progressively shorter faces and wider skulls, to achieve the desired, smushed-face appearance. We may have pushed these dogs very close to the physiological limit, because their lives are now seriously compromised.[13] Brachycephaly squeezes the brain, makes breathing difficult, and is associated with higher than normal levels of disease, especially obstructive airway disease, and early death. In extreme cases, as in bulldogs, the fetal skull is too large with respect to the size of the mother's birth canal, making natural birth extremely difficult and dangerous, often necessitating delivery by cesarean section. Without human-directed breeding, extreme brachycephaly in dogs will likely disappear.

Another striking paedomorphic feature is "puppy dog eyes." Juliane Kaminski and her colleagues compared the skulls of dogs and wolves and found that the facial muscles responsible for raising the inner eyebrows are present in dogs but not wolves. The scientists believe this eyebrow movement "increases paedomorphism and resembles an expression that humans produce when sad, so its production in dogs may trigger a nurturing response in humans."[14] This triggered response in humans is supposedly the release of the "love hormone" called oxytocin.[15]

Kaminski's research illustrates the connection between the evolution of anatomical (facial musculature) and behavioral (motivation to make eye contact) traits. Dogs' motivation to seek eye contact with humans is basically absent in wolves, and so too are "puppy dog eye" muscles. Would puppy dog eyes, or other paedomorphic features for that matter, have any adaptive advantage if there were no humans around? Might they be maladaptive in some way, if humans aren't around to see them and be triggered by them? We won't ever know, but these are exciting questions to ponder and may provide more insights about why these traits evolved in the first place.

An additional, final thought about the anatomy of dogs' eyes relates to a structure called the tapetum lucidum. Many animals who are active in low light have a layer of tissue just behind the retina that reflects visible light back through the retina, brightening their visual world. The tapetum lucidum is what makes an animal's eyes seem to glow green or yellow when caught in a car's headlights. The presence of this structure in dogs' eyes gives us some clues about their natural activity rhythms. The tapetum lucidum might increase the flexibility of dogs' feeding patterns, allowing them to be out at different times. The combination of being dietary generalists and having more time during which they can find food by hunting or scavenging would be advantageous. Activity rhythms, especially feeding patterns, are related to partitioning and sharing of habitat, a topic we'll discuss in the next chapter.

EARS, TAILS, AND COATS

In addition to the primary morphological traits of body size, body shape, and skull shape, other physical traits that have been under selection pressures by humans include the shape and position of ears, the length of tails, and growth patterns and coloration of fur. Although human selection for these traits has been driven primarily by an interest in the physical appearance of dogs, nobody will care what dogs look like in a posthuman world, least of all dogs themselves. Functionality is what will matter.

Like noses, dog ears come in many different and sometimes laughable sizes and shapes. They may be upright, partially upright, floppy, or some combination. Which ear shape and size will be most adaptive is, like nearly every other trait, going to depend on the particulars of where and how dogs are living. Likewise, where and how dogs are living will shape the evolutionary trajectory of dogs' ears in a posthuman future. Looking at wild canids, there is a notable absence of floppy ears because floppy ears are a side effect of domestication. Under natural selection, dogs may eventually all come to have upright ears like their wild cousins.

The primary function of dogs' ears is hearing and using what they hear to make decisions about what's the best thing to do. Dogs on their own will need to hear the rustling of potential prey in the grass or the distant approach of a friend or foe. How acutely dogs will need to hear, and what they need to hear, will depend largely on who or what they need to eat—they only need to hear well enough to find food and avoid being eating. Whether dogs with upright ears hear significantly better than dogs with floppy ears or whether big-eared dogs hear significantly better than small-eared dogs is unknown, but all Transition dogs—except those with hearing loss or deafness—will hear well enough to get along. One important disadvantage of floppy ears, and one reason this phenotype will likely disappear over time, is that floppy ears are more prone to infection than upright ears. Left untreated, chronic ear infections can lead to deafness.[16]

In addition to allowing dogs to hear what's going on around them, ears are used for receiving vocal signals from and sending visual signals to other dogs. It could be that ears of certain sizes and shapes are better tools for communicating than others, and that certain types of dog ears are, as Stephen Spotte suggests in *Societies of Wolves and Free-ranging Dogs*, "too deformed for useful signaling."[17] As with other physical traits, ears present a set of trade-offs and there likely will be variations in future dogs' ear size and shape based on climate, geography, and prey type. For instance, bigger ears might pick up sounds better than smaller ears, but they might also be problematic in very cold temperatures because there is more surface area for heat loss. In cold climates, smaller ears and slightly less acute hearing might be worth the trade-off for protection against heat loss. In warm climates, however, this trade-off may not be necessary. Large ears are not only good for hearing but can also help an animal deal with high temperatures. The strikingly large ears of dogs' cousin, the desert-dwelling fennec fox, offer a clue about ears and habitat. Their ears help them keep cool by radiating body heat.

Moving backwards, let's now focus on dogs' tails. How important are tails to survival and in what ways? As we did with ears, we can look to wild canids for some clues. All wild canids have tails, and our prediction is that tailed dogs would do better overall than tailless dogs and that stabilizing selection would quickly bring back the tail. (Stabilizing selection is selection for a phenotype around a mean, such as a certain color, running speed, or limb length.) As messengers of mood and intention, tails serve important social functions for dogs. Tails can signal aggression, submission, reproductive receptivity, anger, playfulness, calmness, and uncertainty, among other things. Without tails, dogs are missing a key component of social communication, and this may negatively impact how dogs get along with other dogs and how they do within groups or packs. Tailed dogs would also have an advantage in communicating with coyotes and wolves with whom they may mate, or with other canids and individuals of other species with whom they might interact. An additional function of tails is to help dogs balance—watching a tailed dog walk across a log, for example, or jump into the air to catch a Frisbee, you can see the tail being used as a delicate and effective counterweight. One important unknown factor is whether unexpected traits might hitchhike along with the genes for tails, no-tails, or curly tails, and whether these will confer some adaptive benefit.

Dogs' tails, like the rest of their bodies, are covered with fur, and color and coat type have been strongly influenced by the process of domestication.[18] Dog coats come in all sorts of colors, patterns, textures, and lengths. Indeed, one of the most distinctive traits of many established dog breeds is their coat: the spotted dalmatian, the silver Weimaraner, the dreadlocked komondor, the silky-haired Afghan.

While there is wide-ranging diversity in coat color and type in domestic species, there is very little variation in wild types. Some selection factors acting on coats have been eliminated in domestic species such as dogs. As Helmut Hemmer notes, for example, the need for color matching to their environment—for example, for predator defense—acts on wild species.[19] When predator pressure is removed, diversity increases. Likewise, climatic selection for hair type and length acts on wild species, but this selection pressure is removed in domesticated animals and diversity increases. In addition, human-directed selection for certain colors and coat types further increases diversity and, of course, increases the potential for mismatches between coat and context once humans are out of the picture.

We may chuckle at dogs who need to wear puffy coats in the winter or who are dressed up in booties to protect their paws from hot asphalt. But the reality is that many dogs living in the immediate wake of human disappearance will grapple with a mismatch between physical form and habitat, since current human patterns of dog breeding and ownership don't really conform to ecological rules or climatic constraints. Coat type will be one problem. Human environments "zero out" climate extremes. Homed companion dogs are typically provided temperature-controlled shelter that blunts the effects of coat type relative to ambient temperatures. Sheepdogs in Mexico City and greyhounds in Alaska will struggle to survive without human compensation for environmental extremes.

Domestication led to distinct coloration patterns such as tricolor coats, white-socked feet, and brindled coats not typically seen in wild canids. Over time, dogs will lose these markings and become more uniform in color. What color palettes might prevail would depend on predator pressures and ecological contexts. In some ecosystems, dogs might need to change coat color and thickness to adapt to seasonal variation, like the Arctic fox who is white in winter and brownish in spring and summer. There is some evidence that darker pigmented coats are adapted to lower latitudes because the darker pigmentation offers greater protection from UV radiation,[20] and this also could be a variable influencing coat color.

Other "fur-related" traits might make life harder or easier for Transition dogs and dogs of the first few generations. Dogs with long fur that becomes matted without regular grooming may be plagued by skin infections, as are many "wrinkly" breeds such as the shar-peis. Some dogs, such as the Gordon setter, also have what are called "furnishings." These include extremely long and vision-obstructing eyebrows, as found on the bichon frise and shih tzu, and long hair that grows over the face, as found on komondors and other breeds of sheepdog.

We now have a box of clues about how dogs might evolve morphologically once human selection is removed. And we have sketched some ideas about which physical characteristics might be adaptive and which might compromise dogs' potential for survival. But we have a lot more work to do to fill in our picture of posthuman dogs. In the next chapter we'll consider two topics of wide interest for posthuman dogs (and for the humans who theorize about them): food and sex.

4

FOOD AND SEX

Who or what animals eat and who they mate with are the most critical variables influencing whether dogs will survive without us. Without food, of course, individuals will die. Without reproduction, *Canis lupus familiaris* will peter out. Access to food and sex are also the components of dogs' lives over which humans currently exert their strongest and often most self-interested control. Dogs living in human homes, shelters, kennels, and breeding colonies rely completely on humans to determine whether, when, and what they will eat. Although feral and free-ranging dogs are not directly dependent on humans for feeding, many rely almost entirely on anthropogenic food resources such as human waste, garbage dumps, and handouts for survival.

With homed dogs we either restrict them from breeding altogether by surgically de-sexing or socially distancing or isolating them or, alternatively, actively mate them with another dog of our choosing so humans can have their puppies. Reproduction among feral and free-ranging dogs is less well controlled, but humans attempt, in many areas of the world, to stop "uncontrolled" breeding, for example through capture, neuter, and release programs, or through extermination of potential dog mothers and fathers.

In this chapter, we explore how posthuman dogs might satisfy these two basic needs on their own, without human help and without human interference.

FOOD: MEAL PLANS ARE THE CORNERSTONES OF SURVIVAL

When biologists do life-history studies, diet and feeding strategies are among the variables they seek to understand. Acquiring food will be challenging for most posthuman dogs regardless of how they live. Deprived of anthropogenic sources of sustenance, Transition dogs will need to adapt quickly and flexibly to their new situation and find other things to eat. Feeding patterns will evolve over time as dogs become adapted to the demands of the local ecosystems in which they find themselves. Defending food also will be a significant challenge. If an animal can't eat or becomes someone else's meal, the game is up.

Posthuman dogs, like all other wild animals, will need to balance the number of calories they spend with the number of calories they receive in their efforts to stay alive. Many factors could influence dogs' feeding strategies, including anatomical and physiological constraints on what or who dogs can eat, the type of prey available, the distribution of appropriate foods within dogs' home ranges or territories, seasonal variation in food resources, and competition with other animals. Feeding ecology also ties in with dogs' use of space, social behavior and social organization, and reproduction.

Recall that canids, including dogs, are cursorial predators known for their ability to run down prey over long distances. Evolutionarily, canids have shown flexibility in adapting feeding strategies to ecosystem peculiarities, and the diversity of feeding strategies across canids reflects this. Some canids dine almost exclusively on other mammals, whereas others are more omnivorous—they have wide-ranging diets—and along with vertebrate meat they also consume insects or plants. Some canids, such as wolves, are called obligate carnivores because their optimal diet is meat centered.[1] Dogs are omnivorous and will likely be able to survive and even thrive on a broad range of meal plans, from fresh meat to carrion, insects, and plants.

Dogs and other animals also need to drink, and water is as valuable a resource as food and is intimately connected with animals' diets. The presence, absence, and quality of water will be an important factor for posthuman dogs. The availability of water will determine how many dogs can live in a given area and will also determine what food resources there are, since plants, insects, and other animals also require water for their survival.

WHO AND WHAT WILL DOGS EAT?

Let's review what we know now about the diet of current dogs and see how this helps us think about what dogs will do when we're gone. The diet of homed dogs is highly variable and depends on what the owner of the dog decides is most nutritious, cheapest, easiest, or has the cutest packaging. Many homed dogs are fed some sort of kibble or canned foods, designed and processed by dog food manufacturers. These foods generally contain a protein source—often the rendered parts of slaughtered animals that humans find disagreeable, such as the hooves, beaks, noses, and hair—combined with grains, vegetables, or fruits.

As far as researchers know, the diets of free-ranging dogs consist of a combination of anthropogenic food, including garbage; human feces;[2] hand-outs; road kill and other carrion; and small and medium-sized animals caught by hunting, including reptiles, birds, hares, mice, roe deer, and impala. Scientists combing through scats (poop) of free-ranging dogs in Alabama found garbage, grass, leaves, insects, cottontail rabbits, mice, gopher tortoises, and persimmons.[3] Behavioral ecologist Thomas Daniels observed a free-ranging dog scavenging on the remains of other dogs.[4] Dogs have been observed eating very small prey such as mice and, occasionally, larger prey such as kudu and white-tailed deer.[5] Dogs also have been known to kill domestic livestock such as sheep. Although free-ranging dogs appear to hunt successfully at least some of the time, no one knows just how successful they are in relation to how often they try.[6] It is also unclear whether free-ranging dogs hunt cooperatively and, if they do, how frequently. There have been scattered observations of small groups of dogs working together to take down prey, but current dogs seem mainly to acquire food on their own.[7]

FEEDING STRATEGIES: HOW WILL DOGS GO ABOUT GETTING FOOD?

Finding food requires a range of different skills, some physical, some social, and some cognitive. Physical skills will need to be adapted to prey type and to the unique features of different ecosystems. If dogs are trying to catch insects, they will need razor-quick reaction times. If they are trying to chase and catch other mammals, they will need patience, impulse control, speed, and stamina. Depending on what kind of animal they're after, dogs will either need the anaerobic lung capacity and leg strength to run very fast, on and off in short bursts, or will need the aerobic stamina to stalk and follow prey over long distances. Dogs hunting in dense underbrush will need strength to push through foliage, while in an arid and open landscape, speed and agility might be more important.

Dogs also will need physical strength to subdue prey, biting power to kill effectively, and jaw strength to tear flesh from bone. It is possible that dogs no longer have the musculature needed to take down and consume large prey and that smaller mammals, invertebrates, and plants may become their primary food sources. Large dogs may have different strategies for getting food than small dogs. Small terriers who can fit down rabbit holes may have one strategy, while greyhounds who can accelerate and run fast will have another.[8]

Social skills such as clearly communicating intentions about what an individual wants and plans to do, along with the ability to cooperate, will influence whether and how well dogs hunt in groups or packs and make collective decisions, which in turn will inform the size of prey they can catch and how much time an individual can put into getting food. Whether it is best to be alone, in a small group such as a mated pair, or in larger groups including packs will be influenced by local ecological conditions and food availability.

Figuring out how to get what they need will be cognitively challenging as well. The cognitive abilities of dogs will shape what strategies they use for finding or hunting different kinds of prey and how they coordinate efforts among group members. Dogs, as the descendants of wolves, already have the cognitive architecture for predatory behavior. An interesting question about posthuman dogs is whether they will take advantage of hunting-related cognitive patterns that have been under human selection, such as herding (border collies), pointing (German shorthaired pointers), or chasing (greyhounds). Alternatively, the predatory behavior of dogs may have been hijacked by humans in ways that make catching and subduing prey more difficult for them. For example, as one reviewer of this book noted, humans have interrupted the hunting sequence in certain breeds, particularly in sporting dogs such as retrievers who are bred *not* to perform the "kill-consume" segment of the hunt.

Cognitive skills will also shape whether and how dogs share their food with others, and whether and how they store food for future consumption. Some dogs might cache food, which would tap into their ability to remember where food was stored and whether a cache had been exploited and was empty. Red foxes are known to engage in "bookkeeping" by urinating on exploited caches to which they then don't return.[9] Maybe dogs will employ a similar strategy for keeping track of stored food.

There are two sides to the predator-prey coin. Clearly, dogs' prey will need to learn anti-predatory strategies. Although we tend to think of anti-predatory behavior as "run away as fast as you can," there are other strategies prey animals employ to avoid predators, such as altering their own feeding behavior, shifting how and when they move around in areas shared with potential predators, or leaving an area altogether. Shifts in the behavior of prey animals in turn cause changes in plant communities, insects, and microorganisms. We can get a glimpse of these potential impacts by looking at how current homed and free-ranging dogs are already impacting wildlife. In a paper on dogs as predators and trophic regulators (animals with a strong impact on how an ecosystem functions), Euan Ritchie and his colleagues note that the impact of dogs on wildlife goes well beyond simply killing animals that they chase down. Dogs also can instill fear, which can change the behavior, physiology, habitat use, and reproductive success of prey.[10]

As members of the food chain, dogs will not only be predators; they will also be prey and they will need to quickly learn the behavior of those animals who would consider a dog a tasty afternoon snack. Depending on where dogs live, they might be preyed upon by eagles, mountain lions, tigers, wolves, coyotes, hyenas, and other predators. Small dogs might become prey for big dogs. Although dogs, as domestic animals, have been largely shielded from the pressures of potential wild predators, it is likely that they retain anti-predatory behaviors, such as sensitivity to visual and acoustic signals and to chemosensory cues found in urine or feces. Indeed, a study published in *Animal Cognition* found that the odor of potential predators did, in fact, elicit a fear response in dogs, even though the dogs had never been exposed to predation. The researchers found that untrained domestic dogs were able to detect fecal scents from Eurasian brown bears and European lynxes and that the dogs spent less time around the predator scents than they did around the fecal scent of Eurasian beavers—an herbivorous species—or a water control. The dogs also had increased heart rate in the presence of the predator scents.[11]

Many animals, including humans, use mental shortcuts, also called "heuristics" or rules of thumb. These rules of thumb are learned or hardwired ways to streamline cognitive processing and decision-making. For dogs, a rule of thumb might be: "If it's moving in a certain way, give chase," because the odds are pretty good that a moving object is prey. Some variant of this rule of thumb may be why dogs will sometimes begin to chase a plastic bag or leaf being blown by the wind. Another rule of thumb might be: "A bird in the hand is worth two in the bush, so eat what you have," as opposed to dropping a piece of food and giving chase to something fresher or tastier but perhaps elusive.

In one of the only studies on possible cognitive rules of thumb used by dogs, a team studying the feeding ecology of free-ranging dogs in and around Kolkata, India, found that dogs scavenging at garbage dumps preferentially fed on meat rather than carbohydrate-rich waste and used a rule of thumb for getting as much protein as they could from their scavenging expeditions—a rule of thumb the researchers labeled: "If it smells like meat, eat it." Rohan Sarkar, Shubhra Sau, and Anindita Bhadra write, "The dogs were provided with three choices of baskets filled with garbage mixed with either bread pieces (10), chicken pieces (10) or both (5 + 5). The dogs did apply the Rule of Thumb to eat the protein first but did not discard the carbohydrate. They utilized a Sniff-and-Snatch strategy for maximizing meat uptake from the noisy background of garbage. This is a highly efficient scavenging strategy that can help them to maximize food intake, while allowing them to eat the most preferred food first, and has perhaps been adaptive for their survival in a highly heterogenous human-dominated environment."[12]

OPTIMAL FOODS, CALORIC NEEDS, AND TIME AND ENERGY BUDGETS

What and who dogs eat is not necessarily the same as what they *need* to eat to maximize nutritional intake. While there will be individual differences among dogs, taken as a whole, some combination of protein, carbohydrates, fats, minerals, and vitamins will best meet dogs' physiological needs. Although there might be an ideal combination of nutrients in ideal quantities, the more important question for posthuman dogs is what will be necessary for survival where they live and how they will go about getting it.

How much food does a dog need to eat to survive? Most dogs need about 25 to 30 calories per pound per day to maintain their weight. This calculation is based on the average caloric needs of a homed dog and can only serve as a rough guide since the lifestyle of posthuman dogs will be different, particularly in terms of energy expenditure. Using these caloric guidelines, a 30-pound dog will need around 800 calories a day. To put this in some perspective, let's say that a 30-pound dog is trying to survive by eating crickets. If one small cricket is approximately one calorie, our 30-pound pooch would need to catch and eat as many as eight hundred crickets a day. That would be a full day's work and then some. And we haven't even considered how many calories a dog would burn trying to catch these eight hundred crickets.

Ecologists and biologists have done considerable research on how animals balance metabolic needs with energy expenditures. We can pull on a few threads that might give some food for thought on diet and energy budgets of posthuman dogs. Canid biologists David Macdonald, Scott Creel, and Gus Mills divide canids into two groups, those smaller than 20 kg (40 pounds) who eat small prey, and those larger than 20 kg who eat large prey. Tiny prey such as crickets are usually abundant and easy to catch, but they are also weather dependent, as are many invertebrates. The maximum carnivore mass that could be sustained on invertebrates is about 21.5 kg (47 pounds). Getting bigger than 40 pounds requires a different lifestyle, one that includes catching larger prey. In canids, five species have crossed the divide, including wolves.[13] The researchers don't mention domestic dogs, but clearly some dogs will fall into the "eat small prey" category and others will fall into the "eat large prey" category. So, the diets of posthuman dogs are going to diverge significantly depending on their size.

Part of getting enough calories to meet metabolic needs involves balancing the energy demands of hunting with caloric payoffs. But the picture is even more complicated, because competing with other predators is also energetically costly and will almost always be part of the equation. We can go back to MacDonald, Creel, and

Mills's work to see just how costly competition might be. In a study on Kruger National Park's wild dogs (*Lycaon pictus*), the researchers calculated that the wild dogs hunt for about 3.5 hours in a typical 24-hour period and spend the remainder of the time at rest. The dogs have a Daily Energy Expenditure of 15.3 megajoules (MJ) and need about 3.5 kg of ungulate meat daily (4.43 MJ/hour of hunting) to meet their daily energy needs. Hunting is so energetically expensive that if they lose even 25 percent of their kills to spotted hyenas, they must hunt for 12 hours a day, rather than only 3.5 hours.[14] Posthuman dogs may well be faced with the same costs.

SEX: REPRODUCING IS THE KEY TO DOG FUTURES

The loss of human-sourced food will challenge Transition and First-generation dogs and will alter the course of dog evolution in dramatic ways. Just as profound will be the removal of human interference from the realm of dog reproduction. Without humans controlling reproduction, whether by setting up "arranged marriages" and intentional mating between a male and female dog or by disallowing certain dogs to reproduce by spay and neuter practices, dogs will have much greater reproductive freedom. But this freedom will bring with it a range of challenges, from finding willing mating partners to successfully raising young and protecting them from danger.

Before jumping straight into the sack and into topic of sex, let's note that food and sex share a chapter together not only because both are of keen interest to dogs, but also because they will be linked in complex ways for posthuman dogs. Although freedom from human control in the realm of reproduction might be viewed as a boon for dogs, the loss of food subsidies will be a significant challenge for dogs who are trying to make more of themselves. Adequate nutrition is essential to successful reproduction, both in allowing females to carry a pregnancy to term and in the provisioning of pups by mothers and perhaps also fathers after they are born. Females are more vulnerable than males to poor diet, since females bear the energetic burden of pregnancy and lactation. But males also need good nutrition to compete with other males for access to females in estrus and to protect their mates and children.

There are two primary questions related to reproduction: which dogs will be able to survive the transition with good enough fitness and enough energy reserves to successfully gestate, whelp, and raise young—all of which require adequate food resources and shelter from prey and from environmental extremes? And how will reproductive patterns and behaviors shift over time as dogs adapt to a humanless world? Transition dogs will have to quickly learn to fend for themselves—and many won't learn fast enough and will perish. Some will have physical impairments and maladaptive traits resulting from human selective breeding practices that will make the transition difficult if not impossible. There will be a significant winnowing of the dog population. Breeds that have trouble giving birth naturally, such as bulldogs, will go extinct, although the males of these breeds could potentially mate with females of other breeds or mixed heritage. After one or two generations, there are unlikely to be any purebred dogs—unless a group happens to be functionally isolated from other dogs—as humans no longer provide matchmaking services. Moreover, females with no help from males or other "alloparents" will have a very difficult time raising young, thus pushing selection back toward social patterns of reproduction and cooperative breeding found in other canid species.

COURTING, FLIRTING, AND MATING

Let's begin where it all starts, namely, a male and female coming together to make babies. All canids share a general courtship sequence. There is an initial greeting, typically followed by mutual sniffing, circling, and sometimes playing, after which they may get down to business. To take one canid example, wolves forming a pair bond will begin by spending time in each other's company, like teenagers in love. They will press their bodies together, touch noses, gently mouth one another's muzzle, make soft whining noises, and engage in other forms of flirting. Courtship and mating in wolves and other wild canids can extend over days if not weeks.

Observations of mating rituals in free-ranging dogs are sparse but suggest continuity with patterns seen in wild canids. The courtship and mating period for homed dogs is often quite brief, although this is most likely a function of the circumstances under which they interact. In posthuman dogs, the period of courtship may become more prolonged and ritualized, as it is in wild canids.

Without humans playing the dating game for dogs, males and females will have to work hard to attract and then court potential mates. Males will likely try to mate with any female who will have them, while females will be choosier. This is standard Darwinian fare. Females bear the energetic costs of gestation and lactation and will want to make sure their investment is going to be worth it in the long run, that their DNA will be passed on through their children who will then go on to reproduce.

Because there will be competition for mates, mating sequences may be disrupted by competitors who also want to be part of the action. But once copulation starts, the dating game is pretty well complete. When dogs and other canids copulate, the male is essentially locked into the female for a period of several minutes. In wild canids and dogs, competitors will sometimes try to separate a mating pair by breaking the copulatory (or coital) tie, but it's almost impossible.

Which traits might be considered "sexy" to potential suitors? Might it be shiny fur, a long tail, big ears, white patches, big teeth, large "private parts," a seductive odor, or some combination of these? With humans no longer controlling reproduction and deliberately breeding like with like, there would be a rapid loss of "breeds" as we know them. Nevertheless, dogs might preferentially mate with others like them, retaining some general sorting of dogs into similar types. Moreover, there will be some inescapable physical limitations to who can reproduce with whom. An extreme size differential would make the copulatory act difficult or impossible—the size of the vaginal tract and the penis need to be close enough—and would pose challenges during gestation, birth, and suckling.

In natural populations there are mechanisms in place to prevent inbreeding between genetically related individuals. Dispersal is one way by which close relatives move apart, thereby reducing the chances that they will mate and reproduce. Olfactory cues such as "Don't breed with someone who smells like you" also can be a rule of thumb controlling mating preferences. Pack hierarchies based on age and genetic relationship may also play a role in inbreeding avoidance.[15]

Many dogs of the future will likely be mutts, a blending of genetic bits and pieces of the four-hundred-plus established breeds. This may be a beneficial trend for dogs, since a greater mixing of genes tends to be better than less mixing. In addition to the intermixing of genes among diverse dogs, there will also be continued, perhaps even increased, hybridization between dogs and wolves, coyotes, dingoes, and jackals. Dogs already hybridize with these canids in regions where there is substantial overlap in home ranges. For instance, a 2014 study in the Caucasus region between the Black Sea and the Caspian Sea found hybrid activity "in every tenth wolf and every tenth shepherd dog."[16]

The hybridization of dogs with wolves may be influenced by physical traits that resulted from human selection, such as body shape, ears, or tails. To take the example of tails, would Transition dogs without tails, or with tails whose shape might inhibit normal dog communication, such as the extremely curly tail of the spitz or Shiba Inu, be at a significant enough disadvantage that they would be unlikely to survive? Marc remembers a curious tail-related incident in the early 1970s, when some canid researchers were trying to breed a female wolf with a male malamute. The malamute always held his curled tail very high. Around the malamute, the wolf became submissive, tucked her tail, and avoided contact. The researchers then brought in a malamute of the same size who carried his tail in a lower position, and the wolf gladly mated. Similarly, dogs without tails might fail to successfully hybridize with wolves because informational cues are missing. Or perhaps these tail-disadvantaged dogs will compensate by communicating effectively in other ways or may have physical or behavioral traits that blunt the slight disadvantage that taillessness might incur. Transition dogs with cropped tails still have the genes for tails, so if they manage to reproduce their offspring will have all the advantages of a tailed life.

DENNING

For current homed dogs, females generally carry their pups in a safe environment and pups are born in a prearranged "den" that has all the comforts of home. The sperm-producing males often aren't around. Posthuman dogs will not only have to find their own mates, but females will also need to find safe places to bear and raise young.

Digging is a vestigial behavior pattern and one for which dogs are often scolded or punished in human environments. But digging may have important functions for posthuman dogs, not least of which will be creating dens in which to bear and raise young. Studies of free-ranging dogs have included observations of denning. Thomas Daniels, for example, observed communal denning in free-ranging dogs on a Navajo reservation.[17] Sreejani Sen Majumder and her colleagues also observed denning in free-ranging dogs in India and noted that "pet dogs seem to have retained the ancestral habit of denning."[18]

CARING FOR YOUNG

In many domesticated species, including dogs, humans interfere in parenting behavior. Puppies are often taken from their mothers at a young age and are "parented" by people. With companion dogs who are bred and sold, fathers are often completely out of the picture. Mothers still carry, give birth to, and care for the very young. But as soon as puppies are old enough to be taken by humans, weaning and other forms of parental care are interrupted.

Nonetheless, it is unlikely that there have been significant evolutionary changes in canine parenting strategies during domestication. Most parental care is performed by mothers, who provide food, warmth, and protection from predators. Mothers nurse and lick pups, and as pups get older, mothers play with them and help them learn social and other life skills. A mother's care is essential not only to the survival and physical growth and development of pups, but also to their cognitive and behavioral development. In a posthuman future, dog mothers will be allowed to engage in the full natural range of rearing and teaching of pups. If females have more maternal responsibilities and a longer intensive period of mothering, this could have ripple effects. For example, the timing of female estrus cycles might shift from two per year back to one per year as found in wolves.

The lack of data on paternal care in dogs makes it difficult to predict the role of fathers in providing for future dogs. Dogs are the only canids who don't predictably follow a monogamous mating pattern, with mothers and fathers both participating in the care of young. In their observational studies of free-ranging dogs in Italy, Roberto Bonanni and Simona Cafazzo found that puppies were rarely fed by anyone other than the mother. Although some studies report high mortality rates in feral dogs and attribute this to lack of paternal care, Bonanni and Cafazzo note that mortality rates among free-ranging dog pups are like those among wild populations of wolves. So, lack of paternal care may not be maladaptive for dogs.[19]

Moreover, the data on paternal care in free-ranging dogs are not cut and dried. In a small study of free-ranging dogs in West Bengal, India, Sunil Kumar Pal found that four out of eight "copulators" (mated males) stayed with their litters in the first six to eight weeks following birth. When the mothers were absent, these four males were protective of the puppies and acted as "guards," preventing the approach of strangers "by vocalizations or even by physical attacks." One male actually "fed the litter by regurgitation." [20] Pal's findings are intriguing and show that paternal care in free-ranging domestic dogs does occur.

Behavioral biologists Manabi Paul and Anindita Bhadra also found evidence for paternal care in an observational study of fifteen different free-ranging dog groups in West Bengal, India, covering four denning seasons over five years. "Our study," they write, suggests the "highly flexible nature of the breeding system in free-ranging dogs." They found that putative fathers ("putative" because paternity could not be verified) "showed comparable levels of care with the mothers. Mothers invested more effort in nursing and allogrooming, while the putative fathers played and protected more."[21] Dogs of the future might move toward mating systems in which fathers play a more central role, particularly if and where posthuman dogs live in organized groups or packs. Studies of free-ranging dogs have found that males do play at least a limited role in the care of young.

Adult dogs other than parents may also play a role in rearing young, as they do with wild canids. For example, aunts, uncles, or older siblings may engage in helping behavior, or *alloparental* care. Maternal care alone can be enough for the survival and development of pups, but as Paul and Bhadra note, "allocare provides additional benefits."[22] In his research on free-ranging dogs, Sunil Pal and his colleagues observed allomothering that included nursing and regurgitating food. Most of the allomothering was done by matrilineal related females, a pattern also seen in wolves and coyotes.[23]

Research on alloparental behavior in homed dogs is limited, perhaps having been stymied by a longstanding belief that only mothers are involved in caring for pups, but researchers are finally turning attention to the full complexity of dog parenting. In 2019, Hungarian dog researchers Péter Pongrácz and Sára Sztruhala conducted an international survey of dog breeders, asking about interactions of other dogs in the home with a mother and pups. They found that allonursing (nursing by females other than the mother) and feeding of pups by regurgitation were widespread.[24]

Although the research on how free-ranging dogs care for and raise young can certainly provide fodder for hypothesizing about posthuman dogs, there is one critical difference between the situation of current free-ranging dogs and dogs of the future: the presence of human food subsidies. Even where current free-ranging dogs are functionally independent of humans for care, they still almost universally rely on anthropogenic food resources such as garbage dumps. A female dog alone in a humanless future will almost certainly have more limited access to food and will have to expend more energy feeding herself and her pups. As we noted above, the challenges of finding food will undoubtedly push selection in the direction of cooperative breeding. The social rearing of pups and the role of fathers and alloparents will become more pronounced.

FACTS OF LIFE

Many life-history traits are related to patterns of reproduction, including age at sexual maturity, frequency and timing of reproductive cycles, gestation length, number and sex ratio of offspring, and mortality. We have relatively few data on which to base our speculations about how these life-history variables will be balanced in posthuman dogs, but there will surely be some subtle shifts as artificial selection pressures are lifted. Wolves provide some clues into what dog reproduction might look like, particularly in those life-history variables that are more evolutionarily hardwired, such as gestation length.[25]

Female domestic dogs reach sexual maturity during their first year of life, typically at around nine months of age.[26] This is earlier than female wolves, who reach sexual maturity at about twenty-two months.[27] Domestication pressures on dogs have compressed the timeframe for sexual maturity, bringing females into heat earlier and more often, allowing humans to breed more animals and to speed up selection for certain traits. This time frame might decompress under natural selection, with age of sexual maturity moving later for female dogs.

In a wolf pack it is typically the highest-ranking male and female who reproduce. Dominance rank and food availability delay or “suppress” reproduction in lower-ranking animals. In dogs, little is known about the social regulation of reproduction. Homed dogs have few opportunities to form sustained social groupings, so the potential for reproductive suppression simply doesn’t arise.[28] Bonanni and Cafazzo observed packs of free-ranging dogs with multiple breeding individuals, but cautioned that there might nevertheless be some social control over reproduction within the groups.[29] Physiological and behavioral reproductive suppression might occur among female posthuman dogs, and while we don’t know if males also experience physiological reproductive suppression, it’s likely they will be controlled at least behaviorally by other males in their group.

Another reproductive life-history trait is how often females can get pregnant. Nearly all female dogs go through two cycles of receptivity a year, whereas wild canids typically go through one. If humans are out of the picture, posthuman female dogs may go back to one cycle a year like wolves and coyotes. Alternatively, the number of estrus cycles might increase to three, as seen in Carolina dogs, a population of free-roaming dogs in the Southeastern U.S. discovered in the 1970s by ecologist Lehr Brisbin. Carolina dogs are an interesting case because they have been living for years under free-ranging conditions. In a paper on the ecology of primitive dogs, Brisbin and fellow ecologist Thomas Risch wrote that the Carolina dog has traits that have never before been seen in other members of the genus *Canis*. One of the most striking of these traits is that females can have up to three estrus cycles a year, rather than the usual two.[30]

The timing of reproduction is critically important in the wild, where food availability and weather are seasonal. In wild canids, breeding occurs at a particular time of year, timed so that pups arrive when food is relatively abundant. Male wolves show seasonal changes in testosterone levels, with levels peaking during the winter breeding season.[31] Male dogs can and do breed anytime they find a receptive female. We simply don't know if they have a seasonal cycle in which testosterone levels spike and the need to find a mating partner becomes more urgent.

The gestation period for dogs is about sixty-three days, the same as it is for wolves and coyotes. It seems unlikely that this pattern will change in posthuman dogs because gestation length is a very conservative evolutionary trait, meaning that it shows little to no variation among related species and is not strongly correlated with food resources or other ecological variables such as habitat quality.

Litter size and the sex ratio of offspring are key reproductive life-history traits. Both are under strong stabilizing selection (see figure 2.3) and do not vary much across dogs, wolves, and coyotes. Nevertheless, subtle shifts in litter size and sex ratio can occur in response to ecological pressures. In the few studies on litter size in free-ranging dogs, mean litter size hovers around five or six pups. In wolves, average litter size ranges between five and eight pups. Research on wolves suggests that the number of offspring shifts in relation to population density of wolves within an area, with bigger litters born into areas with fewer wolves.

In free-ranging dog litters, sex ratios appear to be slightly male-biased, though in some areas they are female-biased. Sex ratio is relevant to survival because patterns of mortality for males and females are often different. As Daniels and Bekoff noted in their research on population characteristics of dogs on a Navajo reservation in Arizona, potential human intervention makes studying sex ratios and other reproductive traits extremely difficult. Although they observed male bias in the dog litters, they cautioned that females could have been removed from the litters by humans, presumably so that they wouldn't reproduce.[32]

Sex ratio in posthuman dogs may also be dependent on ecological conditions, as it is in litters of wolf pups. In one study,

70 percent of pups were female in an area with low density of wolves; it shifted to 40 to 50 percent female in areas with more wolves.[33] This research was conducted in Belarus, in an area where there were strong hunting pressures on wolves, so this may have influenced the data. Based on his long-term research on wolves in Minnesota, renowned wolf researcher L. David Mech reached a different conclusion: in a saturated, high-density wolf range in northern Minnesota, about two-thirds of the pups in litters were male. Packs from lower density areas had litters with about an equal sex ratio.[34]

Another fact of life is that the reproductive potential of both males and females eventually fizzles out. Dogs, like all animals, have limited time in which to pass on their DNA. Patterns of fecundity (the lifetime ability to produce offspring), longevity, and mortality for posthuman dogs are hard to predict because some strong opposing forces will be in play. The average life expectancy of posthuman dogs will fall somewhere between that of current homed dogs (on the high end) and feral dogs (on the low end). While homed dogs live an average of around thirteen to fifteen years,[35] the survivorship of adult free-ranging dogs is quite poor. Stephen Spotte, for instance, notes that a free-ranging dog of five years "is a true Methuselah," and that most adults probably live three years or less. Village dogs have similarly high mortality if they are not provided any veterinary care.[36] Based on a five-year census of free-ranging dogs in West Bengal, India, Manabi Paul and her colleagues came to similar conclusions. They observed high rates of mortality, with only 19 percent of 364 pups from 95 observed litters surviving until reproductive age.[37]

For current dogs, one of the most significant causes of mortality is humans. Many dogs are killed deliberately—sometimes because they carry rabies, sometimes because they are viewed as dangerous pests, and sometimes because they are viewed as "homeless" (without a designated human home) or as "surplus." Dogs are also unintentionally killed by human activities, particularly by roads and cars. Paul and her colleagues estimated that for free-ranging dogs in India, 63 percent of deaths were caused by humans.[38] For posthuman dogs, the script would be flipped: the disappearance of humans could significantly decrease rates of mortality, but increased survival would be offset by death from starvation once anthropogenic food resources dry up.

Offspring mortality would also likely be quite high for posthuman dogs. Looking again at the sparse data on free-ranging dogs, it appears that only about a third of pups survive the first three months of life.[39] By comparison, pup mortality rates in wolves are somewhere between 40 to 60 percent. Post-natal mortality could be even higher for posthuman dogs if, for example, food acquisition is challenging and females don't have enough energy to provide for their pups. On the other hand, if post-natal mortality in dogs is largely human driven, posthuman pups may have a better chance of survival than their current-day peers.

In the next chapter we move on to consider sociality: how and why dogs might form social groups, how they might cooperate and compete with each other, and what their lives might be like as members of wild communities.

5

FAMILY, FRIEND, AND FOE

As many a dog guardian knows, dogs crave social interaction and will often to go great lengths to assure that they can maintain close contact with their human partner. Given the choice, many homed dogs would join us in bed, eat at the table with us, accompany us to work or school, help us with errands, and go out with us for a night out on the town. Some dogs are so desperate for social contact that they would even shower with their human partner if they were allowed. Dogs seek social contact with other dogs, too. The reason that homes with dogs often have fenced yards is that dogs like to roam their neighborhoods in search of other dogs. On a walk, a dog may pull desperately at the end of the leash for the opportunity to sniff a passing dog's butt. Even dogs' obsession with sniffing—especially spots where another dog has peed—is an important form of social behavior. Without humans as key social attachment figures and without us controlling how and when dogs access each other, the contours of posthuman dogs' social lives will undergo significant changes.

In this chapter we'll explore the ways in which dogs might socially interact with animals with whom they may cooperate, compete, or simply coexist. Sociality is the tendency of animals to form groups and to live together in an organized way. The phrase "social behavior" refers to the suite of interactions that occur when individuals meet, when they come together to form relatively unstructured aggregations or more organized groups, how they interact when they're living together, the ways in which they communicate across space or time, and how they engage in and settle conflicts over the use of space and other resources.

Some posthuman dogs will live in loose groups, some will live in cohesive packs of various sizes, and a good many will likely live alone. Yet all dogs will be social in one way or another. As German ethologist Paul Leyhausen famously argued in his 1965 paper "The Communal Organization of Solitary Animals," no animal is totally asocial.[1] Rather, there are degrees of sociality and a wide range of interactions among individuals can count as social. For example, wolverines live predominantly solitary lives and fall on the low end of the spectrum of sociality. Nevertheless, these solitary animals are still concerned with where others are and what they are doing. When individuals change their behavior in response to where other individuals peed, they are engaging in social behavior. Likewise, if individuals change their travel when they hear a far-off call or signal from another animal, this is also social behavior. And of course, animals necessarily engage in some social behavior when they mate and reproduce. Unlike wolverines, dogs are notoriously social; indeed, they are among the most social of all mammals.

SOCIALIZING PUPPIES

Let's begin with how puppies learn the social skills they need to survive, thrive, and become "card-carrying" dogs. Socialization is the process of acquiring the social skills that will allow an animal to engage in species-typical behaviors. All mammals—domesticated and wild—go through a socialization process. Active socialization begins at birth for dogs, but the most important time for socialization extends from around three to eight weeks of age.[2] This socialization period is often called the "sensitive period." What happens during this time greatly influences how dogs will interact with people, other dogs, and their environment. For humans who keep dogs as pets, the socialization period primarily involves preparing dogs for a life within their households. In the best of all possible worlds, puppies learn to enjoy interactions with people and be well adapted to human environments.

One of the outcomes of "proper" socialization is a dog who is flexible, calm, resilient, and psychologically and emotionally well adjusted. Dogs are encouraged to be comfortable with a wide range of experiences and to be confident when faced with something new or unexpected. A puppy being socialized by a human might, for example, be gently exposed to a wide variety of surfaces, such as grass, concrete, snow, wet pavement, wood floors, or dirt, with the goal being to help the pup establish a level of confidence with their "under-paw" physical environment that will help them remain calm if presented with unfamiliar terrain.[3] These socialization effects should be generalizable to a posthuman world, so that well-socialized Transition dogs will be as prepared as possible for the challenges of a transitioning to a life without human help or friendship.

Will poorly socialized dogs have a harder time transitioning to a life without people than their well-adjusted peers? It's hard to say. Puppies who have been raised in unpredictable environments or who have been exposed to chronic fear or stress, for example by being physically punished for ignoring commands, may as adults suffer from low-level anxiety that prevents them from engaging in exploratory behavior, dealing calmly with novel situations, or taking necessary risks.[4] This might put them at a disadvantage. On the other hand, it is possible that puppies exposed to unpredictable human environments might develop increased resilience, which would be useful in a new and challenging posthuman environment. And different puppies will react differently to the same environment, reminding us of the important role of individual personality in shaping growth and social development in animals. For example, Marc observed distinct personality differences among wild coyote littermates when they first emerged from their den at around three weeks of age. Some pups were bold and adventurous, some were timid, and some could only be described as obnoxious. Similar observations have been made for wolves, arctic foxes, red foxes, and jackals.[5]

A key goal of successful socialization of puppies by humans is to have a dog who is comfortable around other dogs and who understands and responds appropriately to the signals they send in different contexts. This is often achieved through puppy playgroups and careful exposure to other individual dogs and groups of dogs at dog parks or on multiuse trails. Again, if socialization is handled thoughtfully and carefully, dogs usually grow up to have good interpersonal skills in dog-dog interactions. However, many pet dogs wind up "reactive" and behave aggressively toward other dogs. People with reactive dogs often go to great lengths to avoid letting their dog interact with other dogs. What might this reactivity and isolation mean for dogs who are suddenly on their own? Simply put, they might not be able to function well as part of a group or pack.

Transition dogs who began life under feral or free-ranging conditions and who were raised by dog mothers within their natal environment will have been socialized under "natural" conditions. It seems intuitive that dogs raised by dogs will transition to posthuman life more smoothly than dogs raised by humans (though intuitions aren't always borne out). Some dog mothers are better than others, though, and so there will still be variation in how well puppies have been socialized and this variation will certainly influence survival.

After the Transition, pups will be raised by one or both of their parents within dog-only groups and socialization will occur without human intervention. It's interesting to speculate whether posthuman dogs will show the same patterns of development as present-day dogs born into groups that include humans and, if not, how many generations it will take for these patterns to shift to dog-only groups. The assumption of many dog researchers that dogs have evolved a genetic scaffolding specifically directed toward forming affiliative attachments to humans will be put to the test.

Dogs have a more protracted period of socialization than their wild relatives, perhaps because they live in a protected environment and have constant maternal and often paternal care.[6] Over time, the period of socialization might compress because there won't be any humans to help puppies and their parents. Parents and other adults will also have more to do than just raise kids.

Play is part of socialization. Most young canids love to play. Pups of more solitary canid species tend to be less playful than pups of more social canid species. For example, young red foxes and coyotes will often fight before they play, and these "pre-socialization" fights may be very intense and cause injuries or occasionally death. In contrast, young wolves and domestic dogs typically engage in more social play before they fight, and their aggressive encounters are less serious.[7] When posthuman dogs are reared by their mother and other adult dogs, will their playful nature morph into being more assertive or aggressive because they need to mature faster and do things on their own that "homed" dogs didn't have to do, such as take care of and fend for themselves?

Learning is an important component of socialization. Many of the skills dogs will need to survive will have to be taught to them by their mother or by both parents and perhaps other adults. The repertoire of skills dogs will need to survive may shift in a posthuman world, and this might influence developmental pathways and, in turn, the socialization process. For example, there might be skills that dogs would need that could only be acquired as juveniles or adolescents. If dogs were living in a habitat in which they would be feeding on larger-sized prey, they would need to learn methods for stalking and killing various types of prey, requiring that they remain with their parents or natal group until these skills are mastered.

SOCIAL COMMUNICATION: HOW DOGS WILL TALK WITH ONE ANOTHER

Clear communication is profoundly important for the smooth functioning of social interactions between pairs and among individuals living in groups. Our interest here is to zero in on how communication may function in a posthuman dogs' world and how communication patterns might evolve under natural selection. Surely dogs have the communication skills they need to survive on their own. But will there be areas in which they might struggle and what would they be? What might give some dogs an edge over other dogs?

Reading popular dog literature might lead one to think that dogs have evolved to communicate *with us*. We read about puppy dog eyes, gaze-following behavior, oxytocin feedback loops, and even dog ESP. And dogs do communicate with us with great skill. But they also and perhaps primarily communicate with one another when we give them the chance. The communicative repertoire of dogs, wolves, and coyotes is pretty much the same.[8] While dogs may have developed additional communication skills that have emerged through the process of domestication and human-directed selection, they also have the basic tools with which to communicate with one another and many other species.

There isn't much to go on for making predictions about how social communication might further evolve as dogs feralize and once humans aren't controlling their social encounters. One important question is whether dogs' communicative repertoire will change as their bodies change. It is quite possible that some morphological traits or behaviors that may be specifically targeted at communication with humans, such as the musculature that produces "puppy dog eyes," will have absolutely no value in a posthuman world and will fall away. Another possibility is that these structures will be modified and adapted to new circumstances, including an increased need to communicate clearly with dogs and other animals. The unique characteristics of the habitats in which dogs are trying to make a living will also influence how different modes of communication develop.

Dogs communicate with each other and with other animals using a wide variety of signals involving different senses. Current humans who live with and around dogs are often acutely aware of dogs' acoustic signals, particularly barking. Dogs also howl, growl, whimper, and whine. A lot of information about mood and intent can be communicated by a sound or combination of different sounds. Dogs also use vocalizations to communicate their size, saying something like "Don't mess with me—I'm bigger than you." This can be beneficial to dogs on the receiving end of the communication because it can help them judge from a distance or without visual access whether they would be at a disadvantage in an agonistic encounter.[9]

Although dogs share a large acoustic vocabulary with wolves, there are some marked differences in how dogs communicate. In wolves, barking is a minor component of their overall vocal repertoire, used to alert other wolves or to announce, "This is my home or territory." Dogs, in contrast, bark a lot and barking serves a range of purposes in various contexts: defending an area, warning, greeting, playing, and drawing visual attention to the barker. Whether barking will continue to play such a dominant role in posthuman dog communication will depend on whether and under what conditions it is adaptive, maladaptive, or neutral. If barking has evolved primarily under the selection pressures of domestication and is used primarily to communicate with humans, then dogs may eventually bark a lot less.

Dogs also use a wide range of visual signals. They communicate intention and emotion through facial expressions, such as barred teeth or tightened eye muscles. Again, there is considerable overlap in the visual signals used by dogs and wolves. The morphology of certain dog breeds—the shapes of faces, bodies, and tails—may reduce the clarity and range of visual communications. To take one example, a pug can't achieve as much variation in expression with his tightly curled tail or flattened, wrinkly face as a shepherd dog might. It's also possible that dogs with less physical resemblance to wild canids—like pugs—might struggle to communicate with sympatric canids such as wolves and coyotes. This would be particularly relevant if dogs try to hybridize with wolves and coyotes and a lack of understanding impedes mating.

Dogs do a great deal of communicating through scent and olfaction. Like wolves, dogs use olfactory signals—including pee, poop, and pheromones from scent glands in the paws and near the anus—to send messages to other dogs about who they are, where they have been, and how they are feeling. There are ongoing debates about whether dogs use scent marking to acquire and to maintain territories. These discussions have centered on the behavior of homed dogs who may scent mark less than wild canids simply because they aren't given a very broad range of opportunities to use scent and olfaction. Their ability to sniff is constrained by homes and leashes and fences, and they are expected to pee and poop in a narrow range of places chosen by their human. As anyone with a dog knows, domestication has not erased the ability of dogs to scent mark and "read" olfactory signals left by other dogs. Urine marking will be important for posthuman dogs to establish and maintain territories. Posthuman dogs' patterns of marking will probably be like those of wolves. As with other forms of social communication, dogs already follow patterns similar to wolves, so there isn't a whole lot of room for dramatic evolutionary change.

What about the effects of neutering and spaying on scent marking behavior? With respect to olfactory signaling, will Transition dogs who have been reproductively neutralized be at any kind of disadvantage compared to those who are intact? At this point, little is known about how neutering affects olfactory signaling in dogs, so these questions are very difficult to answer. But there have been observations that female dogs may spend less time investigating the urine of castrated male dogs than intact males.[10] Veterinarian and ethologist Ian Dunbar has also observed that castrated male dogs scent mark less frequently than intact males.[11] If desexing does influence communication, for example, by reducing the use of scent marking or changing the chemical composition of urine, then neutered and spayed Transition dogs may be at a disadvantage because the range and subtlety of their communications are diminished relative to their anatomically and hormonally intact peers. This won't be an issue for First and Later Generation dogs.

Dogs also communicate with what are called contact behaviors. Licking, grooming, touching noses, rubbing shoulders together, and sleeping together serve as a kind of social lubricant for communicating affiliative and prosocial—or positive—feelings. These contact behaviors are vitally important to social bonding among parents and young, siblings, friends, and pack members. Contact behaviors can also be important in signaling reconciliation after a fight or other agonistic encounter. There is no reason to think that these contact behaviors are muted or deficient in dogs when compared to their wild relatives, although homed Transition dogs who have had very little contact with other dogs during their lifetime may not have developed a full repertoire of these behaviors or know how to express them.

SOCIAL ORGANIZATION AND SOCIAL DYNAMICS

Dogs clearly form social relationships with humans, and for many dogs a human or family of humans might be the central focus of their social life. But it is far too human-centric to say that the human family is the *natural*, much less the only social grouping of domestic dogs, or that the ur-dog would say to the ur-human (in the voice of Tom Cruise), "You complete me." Dogs also readily form alliances, aggregations, groups, packs, and pair bonds with other dogs. The social relationships among dogs and social dynamics within and between groups of dogs will surely have important survival value.

Social dynamics can be simple or exceedingly complex and can include, among other things, relationships between pairs of dogs all the way up to the organization of a population of animals within an ecosystem and how they aggregate, disperse, and share space. Somewhere in the middle we have the complex dynamics of group or pack formation, where a smaller number of animals try to sort out social relationships among one another. The functioning of a group involves gender dynamics (relationships between members of the same sex and opposite sex); age dynamics (relationships between young and old, young and young, and so forth); social dynamics (play, dominance and submission, appeasement, and leader-follower relationships); and family dynamics (parent-child and sibling relationships). Intergroup dynamics may revolve around dispersal of youngsters from their natal group, dispersal of older individuals who don't fit in, defense of territory, and sharing of or competing over space, food, and other resources.

WILL DOGS FORM PACKS?

One question that elicits great excitement is whether dogs of the future will form packs. A pack is a distinct and stable group of individual animals who hunt, forage, travel, rest, and defend resources together. Packs also engage in cooperative breeding. Packs are usually a mix of family members and outsiders who have been able to join the group. Packs can also form when lone individuals join forces.

There are two questions to answer. The first is, *can* dogs form packs? The answer is undoubtedly yes. Wolves are pack-living animals and the genetic hardwiring for pack living is almost certainly still present in *Canis lupus familiaris*. Indeed, research on free-ranging dogs in different locations shows that they form groups. Sometimes these are loose social "fox-like" groups,[12] sometimes discrete social groups with regular stable membership,[13] and sometimes highly organized packs.[14]

The more interesting questions for us are *whether* and *how* post-human dogs might form packs and under what ecological and social conditions. Group formation will depend on a broad range of variables, including relatedness among individuals, habitat, how much food is available and how it is spatially distributed, the presence of competitors, and the size of potential prey. Within the family Canidae, the species that form packs tend to be larger bodied, such as wolves, African wild dogs, and dholes, a canid native to central, south, east, and southeast Asia (also called the Asiatic wild dog). Smaller canids tend to be more solitary, although red foxes have been observed living in groups. Group formation in larger canids is associated with a reliance on larger prey, which in turn provides an incentive for cooperative hunting. Following this logic, would large dogs tend to group together and form packs, while small dogs remain more solitary?

We can gather ideas about what future dog packs might look like and how they might function by looking at homed as well as free-ranging and feral dogs. Dogs who live in human homes obviously don't form packs, and even the formation of social groups is rarely observed. This is because the conditions under which these dogs live and under which we observe them place severe constraints on the formation of lasting social groups. Homed dogs are often forced into behavior patterns suited to a more solitary animal than a highly social one, having limited opportunities to coordinate their actions with potential companions or competitors. They may be less motivated to form social groups with other dogs if they are highly socialized to humans and if their basic needs for food and shelter are met. Their attempts to form groups—for example, at a dog park—are often awkward and unsuccessful, perhaps because the conditions for group formation are unstable and fleeting and because humans interfere. The fact that humans don't observe homed dogs forming packs doesn't mean that dogs are behaviorally incapable of pack formation, but simply that the right social and ecological conditions have not been met.

Data on free-ranging and feral dogs offer a more robust, though ambiguous, set of clues. Early research on free-ranging dogs seemed to suggest that they are largely solitary or form only small and temporary associations with one another. However, a more complex picture has emerged over the past two decades. Free-ranging dogs have been observed forming stable groups, usually consisting of between two and twelve individuals, with most groupings falling on the smaller end of this spectrum. Although some groups of dogs appear stable, in his work on free-ranging dogs in Baltimore, Maryland, Alan Beck observed that groups seemed to form and dissolve very quickly, often within days or even minutes.[15] In other studies of free-ranging dogs using direct observations and radio tracking, individuals in some groups have been observed cooperating in territorial defense. Defense of territory does not always involve direct confrontations between groups, but rather large groups are simply more conspicuous, and this might reduce the likelihood of fighting.[16] It's also possible that in some locations, dogs don't want to be visible so individuals may choose to travel on their own or in smaller groups.[17]

Although the line between free-ranging and feral is gray, it may be that differences in social behavior emerge as dogs become wilder. Research by Daniels and Bekoff suggests that truly feral dogs may have different patterns of social organization than free-ranging dogs. They noted that free-ranging dogs in both urban and rural settings were less social than feral dogs, who characteristically live in packs.[18] They also observed seasonal variations in feral dog packs, suggesting that reproduction influences the social structure of packs both directly by pups being born into the pack and indirectly by pregnant females temporarily emigrating from the pack to give birth.[19]

GROUP COMPOSITION: WHO WILL "PACK-UP" WITH WHOM

What might be the composition of these posthuman dog packs? At least in the beginning, as groups of dogs are forming, they will be made up of unrelated individuals since dog families are unnaturally and often broadly dispersed by humans. Over time, packs of posthuman dogs will probably resemble those of wolves.

Packs of wild canids are typically composed of individuals of roughly the same body size and morphology. Dogs may form packs with other dogs of similar body size, so that there will be little-dog packs and big-dog packs. Alternatively, packs composed of large and small dogs could reap the benefits of group living, while also not directly competing for precisely the same food resources. Successful groups tend to be made up of individuals with different personalities, higher and lower rank, and leaders and followers. These differences in behavior, sometimes called behavioral polymorphisms, can help maintain the integrity and cohesiveness of a group and can influence dispersal patterns.[20] A group made up of all dominant personalities or leaders, for example, would not function smoothly. Following on this logic, dogs from different breeds with different temperaments might come together to form successful groups.

Another speculative question is whether dogs from vastly different experiential backgrounds (homed dogs and feral dogs) would be able to come together to form successful groups. One of the reviewers of this book said they most likely would, based on her observations of free-ranging dogs around Rome. In Rome, house dogs who were abandoned would often gradually be accepted within already existing packs of free-ranging dogs. Sometimes, abandoned pet dogs would group together and form their own small packs, usually with three or four individuals. Some dogs were "eventually driven out, attacked and sometimes seriously injured." Females appeared to be more easily accepted than males and certain breeds of dog fared better than others. "The less communicative breeds," she noted, "undoubtedly had more problems in adapting with free-ranging dogs living in the area."[21]

RESOLVING DIFFERENCES OF OPINION WITHIN GROUPS

Individuals and groups need to be able to deal effectively with conflict. Conflict between lone and paired dogs, within larger groups of dogs, and among groups of dogs will certainly be a fact of life and may even be intense, particularly if resources are limited. Whether the process of domestication has reduced the abilities of dogs to avoid and resolve conflict is unclear. German ethologist Dorit Feddersen-Petersen argues that the social behavior of dogs has changed significantly from wolves and that dogs have lost some of wolves' impeccable interpersonal skills. Although her conclusions are based on limited data, they raise interesting questions about the effects of selective breeding on dog behavior. "In our research," she wrote, "we found that some dog breeds are unable to cooperate (on a very basic manner: just doing things together) and compete in groups, reflected in difficulties in establishing and maintaining a rank order." Poodles, according to Feddersen-Petersen, are particularly inept. She went on, "The interactions in these dog groups are not functional, and the members have difficulties coping with challenges from the environment. It is striking that tactical variants of conflict solving (to appease, animate or inhibit the opponent), a common practice in wolves, do not exist in groups of several dog breeds. . . . Within many groups of dogs, trivial conflicts often escalate into damaging fights."[22]

Italian dog researchers Roberto Bonanni and Simona Cafazzo come to a different conclusion in their research on pack stability in free-ranging dogs. "Our results suggest . . . that evolution in a domestic environment has not substantially altered the ability of dogs to form structured packs with conspecifics."[23] One explanation for the discrepancies between Feddersen-Petersen's and Bonanni and Cafazzo's conclusions is that Feddersen-Petersen was studying purebred pet dogs, while Bonanni and Cafazzo were studying free-ranging dogs, the majority of whom were likely mutts.

Structured packs generally have conventions of rank or hierarchy. This is true of wolf packs and will also be true in dog packs. Although hierarchies impose certain costs on individual animals, such as risks of being last to get food or not being allowed to reproduce, the benefits of living in a well-functioning group typically outweigh the costs. Hierarchies help animals maintain cohesive groups, particularly by providing conventions that reduce the likelihood of conflict. When a hierarchy of dominant and subordinate individuals has been established, there is no need to fight each time competition over resources arises because everyone knows where they stand and how resources will be distributed. Fighting is risky and energetically costly, so everyone benefits from fewer fights. Bonanni and Cafazzo note that social dominance is a convention that allows animals to resolve social conflicts in a relatively peaceful manner. They also note that dominance can be expressed in the context of affiliative relationships, not just agonistic ones.[24] As German ethologist Rudolf Schenkel observed, submission in wolves and dogs is "the effort of the inferior to attain friendly or harmonic social integration."[25] Conventions of rank help keep the peace.

MOVING ABOUT

The ways in which animals use space—whether they stay put, travel, mark or defend territory, and share resources within a given geographic area—is a key topic of interest for biologists trying to understand social behavior. Although moving about may seem like something an individual animal does, use of space is actually a highly social activity. It involves complex communication and negotiations and among groups of animals about how to divide up, share, and compete over spatial resources. When biologists write about space use by animals, they rely on the concepts of *home range*, *core area*, and *territory*. Home range was defined in a classic paper by William Burt as "that area traversed by the individual in its normal activities of food gathering, mating, and caring for young."[26] Animals use home ranges for the basics of survival, such as raising young, provisioning food, traveling, and sheltering from predators and the elements. A core area is a smaller sector within a home range, in which an individual spends roughly half of their time. An animal's territory is a subsection of a home range over which an animal has exclusive or primary use. Territories are actively defended against intruders of the same and other species. The defense of territory may be physical (tooth and claw) or may be achieved indirectly through olfactory signals, such as scent marking; visual displays, such as bared teeth and raised hackles and tail; or auditory signals, such as barking and howling. Individuals and groups are said to be territorial when they defend these areas.

Home ranges, core areas, and territories will be of vital importance to posthuman dogs. What these might look like and how exactly dogs will negotiate the partitioning of space with one another and with other species remain open to considerable speculation. It is hard to know how space-use behaviors in dogs may have been shaped by the domestication process and how these behaviors will be shaped by the different ecological contexts in which dogs grow up and in which they live.

Space use by homed dogs is poorly understood and extremely difficult to study. Humans exert so much control over where and when homed dogs are allowed to move around that the data—if there were any—wouldn't be very useful for speculating about posthuman dogs. The use of space is one of the better researched areas in feral and free-ranging dog behavior, and although there is much that we still don't understand, we can nevertheless gather from this some strong clues about home range size, territorial behavior of dogs, and possible relationships between territoriality and ecological variables such as food availability.

In 1973, Alan Beck published some of the earliest observational studies of space use by free-ranging dogs in Baltimore, Maryland.[27] The roving dogs Beck studied had an average home range of 0.26km^2. Stephen Spotte's *Societies of Wolves and Free-ranging Dogs* provides a comprehensive review of the literature on the home range sizes of free-ranging dogs. Spotte reported significant variations, although one pattern that seemed consistent was that home ranges in urban areas were far smaller than those in rural areas, with urban and rural indicating the density of human populations rather than the density of dog populations. In urban areas, home ranges tended to be less than 10 hectares (0.1km^2). Food resources in urban areas are concentrated and relatively abundant because of the amount of human waste and garbage. In rural areas where food resources are more spread out, home ranges were often considerably larger: studies found home range sizes between 0.2 hectares (0.002km^2) and 2,850 hectares (28.5km^2).[28] Other researchers have also compiled data on home range sizes in free-ranging and feral dogs, and they too noted extremely large variations depending on where dogs were living.[29]

It's difficult to use existing data to project home range size of posthuman dogs because the biomass of food without humans will be drastically reduced and differently spaced. But we might predict that home range size will mirror patterns observed in rural dogs with food resources being widely dispersed.

As suggested by available research, home range size and territoriality are dependent on what type of food is available and where it's located. Matthew Gompper, in his book *Free-Ranging Dogs and Wildlife Conservation*, described behavioral differences between free-ranging dogs living in Italian villages and those living in an adjacent but more rural area. Village dogs were more solitary than rural dogs, who lived in territorial packs. Both groups relied heavily on human-derived foods but developed different forms of social organization and different feeding strategies.[30]

Posthuman dogs will also have to defend their food resources, and different foraging ecologies will lead to variations in social organization and territorial defense. Bonanni and Cafazzo hypothesize that since free-ranging dogs subsist on abundant food resources and have higher density populations than their wild ancestors, they may be less territorial than wolves. Between-group territoriality in free-ranging dogs appears to be influenced by multiple factors: density, abundance, predictability, and distribution of food resources.[31] It is also possible that dogs of different sizes could occupy the same home range since their preferred foods would be different enough that there would be little competition over resources.

Several studies have found that space-use patterns vary seasonally and when there are dependent pups. Thomas Daniels, for example, observed seasonal patterns in several feral dog packs living on the Navajo reservation that spans parts of Arizona, New Mexico, and Utah. Adult dogs restricted their movements to a smaller area when pups were young, a common canid pattern.

When pups became more mobile and independent, adult dogs expanded their range. "In the Canyon Pack, once pups were able to survive without adult care, beginning at about 4 months of age, the mean home range increased more than 10 fold."[32]

As previously noted, free-ranging dogs use acoustic, olfactory, and visual signals to regulate space use. For example, Daniels and Bekoff describe barking as "an unambiguous signal indicating the resident's intention to contest an intruder."[33] Scent marking with urine, feces, or glandular secretions can create a "fence" and signal territorial intentions, although the fence is transitory and needs to be frequently refreshed. Visual territorial signals include postures such as a stiff gait and high tail, and facial expressions such as bared teeth. These clearly tell potential intruders that "this is my home, not yours."

Several other behavioral patterns might also inform our predictions about use of space by posthuman dogs. Home range size in mammals tends to scale with body mass. In other words, home range size gets larger as animals get larger. Can we use this general mammalian pattern to make predictions about how large an area might constitute a home range or territory for dogs? Wolf territories range from 50 square miles to more than 1,000 square miles. The territories of feral dogs are generally much smaller than those of wolves, and much smaller than might be expected, given their average body size. It is possible that home ranges and territories will become larger for posthuman dogs in the distant future.

Another mammalian pattern is called natal philopatry (a "love of home"), the tendency of an animal to stay within or regularly return to a certain place, often their birthplace. Remaining in one's familiar home range is generally less costly energetically than moving to a new place. Some posthuman dogs will certainly remain in place. However, some will find themselves in habitats that are hostile, barren, or overpopulated and will be forced to move. We don't know what variables are most important for the successful colonization of new areas by dogs.

NEIGHBORLY RELATIONSHIPS

Homed dogs learn to share space with their humans with varying levels of success, as anyone who has ever tried to get a good night's sleep alongside a "bed hog" dog can attest. When dogs are on their own, they'll be sharing space not with humans, but with a whole range of local wild species with whom they will *cooperate*, *compete*, or *coexist*—what we call the "three C's." Sympatric relationships—or relationships among species who share the same geographical area—can vary over time and place, depending on the availability of food and other resources. Dogs may be interacting with other canids, such as foxes, coyotes, wolves, dingoes, jackals, and African wild dogs. In formerly urban areas dogs may share space with animals who have adapted to living in and nearby human settlements, such as cats, raccoons, deer, rats, mice, and various carcass-consuming and garbage-eating birds. Dogs in rural or wilder areas may encounter a different, though perhaps overlapping, cohort of animals.

With humans out of the picture, dogs are going to change the landscape of the three C's. For example, in areas where dogs, wolves, and coyotes share space, there could be shifts in intraguild competition, or competition between sympatric carnivores.[34] (A guild is a group of species that exploits the same resources in a similar way.) Wolves, coyotes, and dogs have similar meal plans, so in areas where all three species are present, they might compete for the same basic pool of food resources.[35] But will dogs compete with wolves over large prey, such as elk? Or with coyotes over smaller prey, such as mice and rabbits?

There may also be intraguild predation, meaning that competing carnivores prey on one another and not just on similar prey. So, where wolves, dogs, and mountain lions live in the same space and compete over food resources, dogs may also *become* part of the meal plan for a wolf or mountain lion, just as a wolf or mountain lion could potentially become prey for a dog (although, admittedly, this stretches the imagination). Ecologist and conservation biologist Abi Vanak and his colleagues note that dogs may compete with a wide range of other carnivores, including vultures, eagles, marsupials, civet cats, badgers, lions, and hyenas, to name just a few of their potential neighbors. The intensity of competition will be influenced by dogs' relative position within the native carnivore community, the density of dog populations, and the tendency of dogs to form packs.[36]

As an example of how sympatric relationships of dogs depend on their size, behavior, and feeding ecology, consider the relationships between dogs and wolves. Wolves and dogs might cooperate, compete, or coexist with one another depending on ecological conditions. On occasion they might even breed. Smaller dogs may not be viewed as competitors by wolves, and they might simply coexist with them. Alternatively, smaller dogs might be viewed as potential prey rather than potential competitors, in which case none of the three C's would come into play, at least from the wolves' point of view. Larger dogs would more likely be viewed as competitors, especially if they are going after the same prey as wolves. However, if dogs hunted and scavenged only small prey, insects, and plants, wolves might coexist beside medium- and large-sized dogs with minimal conflict.

Wolves and coyotes are sympatric species whose interactions are of interest in understanding how sympatric relationships might unfold as ecological systems shift and animals move around. When wolves were reintroduced into Yellowstone National Park in 1995, after having been wiped out by humans in 1926, they rebounded quickly. But as wolf packs formed in the Lamar Valley, they decimated the coyotes living there. The Lamar Valley was a closed system as far as food was concerned, which is one possible explanation for the intense competition between wolves and coyotes. If humans were to disappear, there might be numerous "dog reintroduction" experiments as dogs redecorate nature, become wildlife, and begin to disperse into ecosystems as wild animals, instead of as pets.[37] The more limited the food resources in ecosystems into which posthuman dogs trespass, the more likely that competition with sympatric species will result, rather than cooperation or coexistence.

The range of relationships that dogs might have with sympatric species is enormous and may not fit easily into either the predator-prey binary or our three C's. A fascinating study by ecologist Erin Boydston and her colleagues on dog-coyote interactions found that social interactions fell on a continuum between playful and agonistic. Dogs and coyotes both initiated play with one another, and dog size seemed to be related to types of interactions that followed. While large dogs were sometimes the targets of agonistic behavior from coyotes and vice versa, no small dogs were involved in agonistic encounters.[38]

One highly speculative explanation for Boydston's observations is that wild animals have schema, or cognitive shortcuts, for recognizing other animals with whom they share ecosystems. Perhaps large dogs fit the schema for coyotes, while small dogs do not. Broadening this to other situations where posthuman dogs are trying to integrate into a particular ecosystem, some dogs might adequately fit a schema for other wild animals in the area (such as dogs who resemble wolves or dingoes), while some dogs (pugs or

French bulldogs) would be too different from what a wild animal might recognize.

To step back, we've talked about how dogs might look, what or who they might eat and how they will reproduce, and how they might manage social relationships with each other and with other animals with whom they share space and resources. We move on now to what might be the most compelling question about post-human dogs: who they will be, on the inside, without *Homo sapiens*, their key partner in evolution.

The reason people love dogs so much is not solely because of the way they look on the outside. It isn't necessarily their puppy dog eyes, their cute floppy ears, or their soft fur. It is also who dogs are on the inside that truly draws us to them and that provides fertile soil for the deep friendships and mutual loyalties we share with them. In the next chapter on possible evolutionary trajectories of posthuman dogs we will explore their inner lives.

6

THE INNER LIVES OF POSTHUMAN DOGS

As anyone who lives with a dog knows, and as the hundreds and hundreds of books and research papers on dog cognition and emotion confirm, these animals are whip-smart and keenly perceptive. Dogs often seem to know what we are thinking or feeling even before we do. Some of the canine capacities that might be most relevant to their future survival include the ability to rapidly process information, learn new skills, solve problems without succumbing to frustration, assess risk, and accurately read the intentions and emotions of others. But which of these cognitive and emotional traits will be particularly relevant to posthuman dogs, and how they will be used and modified in new contexts? And how will these traits evolve in response to the new challenges? Cognitive and emotional capacities of dogs—what they know and what they feel—will shape their survival in a posthuman world and will, in turn, be shaped by the kinds of challenges that dogs face.

As we've seen in previous chapters, initial intuitions about which traits or behaviors will be most favorable often deserve at least a second look, which is why this imaginative project is so interesting.

THINKING AND KNOWING

Just as there is tremendous diversity in the morphological traits of dogs, so too are there marked variations in individual cognitive capacities. Simply put, "cognition" refers to mental processes such as perception, learning, attention, working memory, long-term memory, decision-making, problem-solving, and intelligence.[1] Recall from our earlier discussion that while certain physical characteristics define dogs as a species (for example, four legged carnivores with one tail, two ears, and a highly efficient nose), there are also marked individual variations. Every dog has a tail, but some tails are shorter than others, some furrier, and some straighter. Likewise, all dogs share certain sensory capacities that help them make decisions, including an acute sense of smell and very good hearing, good visual acuity in low light conditions, communication through olfactory, visual, and auditory signals, and observational learning.[2] Yet each dog has a unique cognitive profile. Variables can combine in an almost infinite number of ways, and an individual dog who may be at a disadvantage in some respects may have other capacities that tip the balance back in their favor. A dog might be relatively unskilled at reading the emotions of other dogs, for example, but be adept at hunting squirrels.

BEHAVIORAL FLEXIBILITY

Donald Griffin, the "father of cognitive ethology," stressed that behavioral flexibility, or plasticity—the ability to make a wide variety of adaptive decisions when faced with changing conditions—is a marker of consciousness. There is now a large body of research within the fields of ecology and cognitive science on how animals modify their "cognitive performance" in response to changing and novel environmental challenges and how plasticity in cognitive abilities provides evolutionary advantages that promote survival.[3] Dogs have fairly high levels of behavioral flexibility; there is also significant individual variation among them.

Individual behavioral flexibility will most certainly be key to the survival of dogs in a posthuman world. In all future scenarios there will be significant changes to the ecological challenges faced by dogs, from the abrupt loss of human food subsidies to the newly valuable skills of being able to successfully navigate social interactions with other dogs and other animals. Some dogs will weather these challenges better than others because they'll be more flexible and will be able to come up with workable solutions to new social and ecological problems.

LEARNING

Dogs excel at learning. But how and what dogs learn from others will shift dramatically in a posthuman future. Dogs will no longer learn from humans, nor will dogs need to learn how to understand and interact with humans or in human-dominated environments. However, they will need to learn to respond to novel challenges, and they'll need to learn fast.

Let's begin with homed dogs, who perhaps learn the most from and about humans. Much of their learning is directed at figuring out how to navigate the unique ecological niche of a human home and some of what they learn is actively taught to them by humans. Dogs are "trained" to respond to certain verbal or gestural cues, to stay a certain distance from a human when "on a walk," and to avoid certain places in their home ecosystem, for example, a couch. While homed dogs may learn a wide range of skills and "commands," it is unclear how many of these will be useful in posthuman settings. Knowing tricks such as "come, sit, stay" or "spin for a treat" will be irrelevant to dogs on their own, but the impulse control dogs necessarily learn as a by-product of human training may come in handy.

Still, homed dogs know a lot more than what they have been directly taught by their human family. Indeed, perhaps most learning that takes place among homed pet dogs is not imparted on purpose by a human but rather is self-directed by dogs. Dogs are keen observers of human behavior and learn through trial and error how best to shape their own responses to get what they need from us or to avoid unpleasant experiences. Some dogs must learn how to adapt to a rapid succession of very different ecological niches, from a home to a shelter to a different home. A large number of dogs navigate these complex challenges successfully.

Comparisons between dogs and wolves may shed light on how the process of domestication has shaped the genetic predisposition of dogs to learn. One of the most concentrated areas of research has been how dogs learn to follow human communicative cues, particularly "distal cues" such as a finger pointing in the direction of what we want a dog to see. Not surprisingly, homed dogs follow pointing gestures more reliably than free-ranging dogs, and homed and free-ranging dogs both do better than wolves.[4]

Little is known about learning and transmission of knowledge in free-ranging populations of dogs. We don't know much, for example, about the transmission of social and cultural knowledge from older generations to younger individuals within or between groups of dogs. A greater understanding of free-ranging dog cognition can perhaps help us untangle which aspects of learning are genetic and which are due to human socialization.

PROBLEM-SOLVING

As individuals and as groups, posthuman dogs will be faced with numerous novel problems. These include finding and keeping food; finding shelter; establishing and protecting homes ranges and territories; forming alliances; developing and maintaining pair bonds; and resolving conflicts.

Dogs who are more behaviorally flexible won't be locked into trying to solve a problem in one or only a few ways. Some other characteristics that may play a role in problem-solving are persistence, curiosity, seeking out and responding to novelty (neophilia), responding to frustration, risk-aversion, and openness to help from others. Variation among individuals in problem-solving capacities will benefit dogs living in groups or packs because individuals will complement one another in what they know, how they learn, and how they identify and solve problems.

Which problem-solving skills might be most relevant for posthuman dogs? It will depend on what kinds of problems dogs are trying to solve and the context within which they are solving them. Dogs who learn to solve problems rapidly will be at an advantage, especially when it comes to life-sustaining skills, such as hunting and recognizing danger. Nevertheless, dogs who take their time to think through a problem before jumping to a conclusion may have an edge in certain situations. Some individuals will be able to deal with many different variables simultaneously or sequentially, whereas others may do better when they can focus on a single situation or set of stimuli rather than a mixture of different sorts of input. In other words, some will be good multitaskers, whereas others may do better in more focused contexts. Being aware of what they don't know, an aspect of metacognition, might also be a valuable problem-solving skill for dogs, since they might be willing to give up on a task and go on to something else when it isn't worth the energy.[5]

Another aspect of successful problem-solving is persistence in "object manipulation"—figuring out what a novel object is and what it can be used for. In a 2019 study on persistence in captive and free-ranging dogs, dog researcher Martina Lazzaroni and her colleagues compared how dogs with different life experiences responded when presented with a novel object containing food. They found that homed and captive dogs were more persistent than free-ranging dogs in manipulating the object to try to get at the food reward.[6]

Persistence may be associated with success at problem-solving, but only up to a point. When persistence bleeds into perseveration, problem-solving capacities diminish. In a 2012 study on problem-solving in spotted hyenas, biologists Sarah Benson-Amram and Kay Holekamp found that although persistence and diversity of behavioral responses to a novel problem were associated with higher levels of success overall, the most persistent animals were not actually the most successful at solving a puzzle problem. The most persistent hyenas sometimes got into a rut, trying the same thing over and over rather than thinking outside the box. Benson-Amram and Holekamp write, "Perseverative errors occur when individuals repeat the same behavioural response over and over, despite the absence of any stimulus or reward, and such perseveration is thought to inhibit problem-solving and learning. To solve problems reliably, individuals must avoid such errors and instead seek out alternative solutions to the problem." They conclude, "Our results suggest that the diversity of initial exploratory behaviours, akin to some measures of human creativity, is an important, but largely overlooked, determinant of problem-solving success in non-human animals."[7] Like the hyenas, posthuman dogs will also face novel food acquisition "puzzles," and innovation, creativity, and flexibility—along with a healthy dose (but not an overdose) of persistence—will be key to the success of dogs.

Based on research into life experiences and persistence, we might imagine that Transition dogs, particularly those with a lot of human contact, will approach problems differently from First-generation and Later-generation dogs. But there are many open questions. For example, would highly trained homed dogs be any better at solving problems than dogs who had no "professional" (human-taught) problem-solving education? And if so, would these problem-solving skills translate into survival? On the one hand, it might be that highly trained dogs would be locked into a more specific and narrower set of responses, like the perseverating hyenas, which would be detrimental when on their own. On the other hand, these trained dogs may have learned that if they try something and don't get rewarded, they should change their behavior and try something else, a behavioral pattern that might translate into greater adaptability. A few studies have found that in homed dogs, training improved dogs' overall problem-solving success.[8]

Another open question is whether dogs who are particularly "smart" or skilled at something would have an edge in posthuman futures. Would a dog like the famous border collie Chaser, who could understand one thousand words, have a better chance at survival than Chester, who only knows one word ("sit")? It is hard to say, but the question invites some interesting lines of speculation. Chaser was very good at learning words and making inferences, but would his world-class vocabulary-building talent have been generalizable to other kinds of problem-solving? Chester may simply have been unmotivated to solve the sorts of problems presented by his human owner (maybe Chester found vocabulary building boring) and may do better than Chaser when faced with problems of finding food or attracting a mate.

Problems also are often social in nature—either they are problems of social interaction, such as resolving conflicts; negotiating the peaceful distribution of resources, such as space; or figuring out what another dog is thinking and feeling. Dogs are highly skilled social problem-solvers. Even at dog parks, which often involve chaotic collections of dogs who are meeting for the first time in a relatively high-arousal environment, dogs manage to recognize and resolve social problems, and fights are surprisingly rare.

Dogs will work together to form cohesive groups and to accomplish shared goals. If you watch groups of dogs forming at a dog park, you'll often see that movements are remarkably coordinated—that dogs engage in dances of shared intentions they choreograph almost instantaneously on the run, even in groups of dogs who are meeting one another for the first time. Anyone who has lived with more than one dog knows that they coordinate their activities. Still, the research on how dogs solve problems requiring social coordination or cooperation is young and there is much we still don't understand.

Studies of cooperation in animals are often conducted under highly controlled conditions in captivity. The study question often takes the form "What can we get animals to do together to accomplish a shared goal?" One of the most common of these "shared goal" tests is a rope-pulling task. In this experimental set-up, two animals must simultaneously pull on a rope to move a food-laden platform into eating range. The rope-pulling task has been used on a wide range of species and in various iterations to compare the tendency to cooperate. Dogs and wolves have been of particular interest to researchers. Sarah Marshall-Pescini and her colleagues concluded that wolves seem to cooperate with one another better than dogs and that wolves and dogs will engage in interspecific cooperation.[9] Friederike Range and her research team found that both wolves and dogs will recruit humans to help with a rope-pulling task.[10]

These experimental results help us understand the cognitive underpinnings of behavior, and the experimental setups allow us to control variables and get more robust answers to specific questions. Whether or not these experimental results then apply back in the messy real world is unclear. More uncertain still is how these results might apply to posthuman dogs. Do the data tell us that dogs aren't as skilled at cooperating as wolves and might not easily form cohesive packs and will thus lose out on all the potential benefits of pack-living? Or is it that captive dogs simply haven't had life experiences involving the need to cooperate with others?

A series of rope-pulling experiments involving captive spotted hyenas adds additional details to our understanding of how social and individual factors influence complex cooperative problem-solving that might well apply to free-ranging dogs. Animal behavior researcher Christine Drea and her colleague Allisa Carter found evidence that hyenas engaged in complex cooperative behaviors to achieve a goal. Pairs of hyenas synchronized and coordinated their behavior during cooperative problem-solving and made behavioral adjustments in response to their cooperating partners. The animals, Drea and Carter conclude, "coordinated their behaviour temporally and spatially to solve cooperation tasks that modelled group-hunting strategies."[11] Because hyenas and dogs are both social carnivores, these results are perhaps suggestive in our consideration of how posthuman dogs might work together to solve problems such as acquiring food.

One problem that social animals face is "interpersonal" conflict. In chapter 5 we discussed the role of rank and affiliation in social dynamics. Problem-solving skills are another key element in the social dynamics of a group or dyad. The rope-pulling research by Marshall-Pescini and her colleagues offers some clues about how dogs solve problems related to social conflict. In their discussion of why wolves might have outperformed dogs on the cooperation task, the researchers note that for the dogs, conflict-avoidance may have been a priority.[12] We certainly don't have a complete picture of how dogs solve interpersonal problems or how they work together to solve problems, and further research on cooperation and coordination in free-ranging dogs could add additional details to speculations about posthuman dogs.

LINKING THINKING TO ECOLOGICAL CHALLENGES

In his 1995 book *The Thinking Ape: The Evolutionary Origins of Intelligence*, Richard Byrne explored the interactions between environmental demands and a species' cognitive skills. Byrne examined how variations in food availability led to differences in a whole range of behaviors and sensory capacities in various primates and other animals. In one of his most well-known examples, Byrne describes the evolution of advanced spatial skills in frugivorous primates who need to know when and where different fruits bloom.

As Byrne's work showed, animals evolve a certain set of cognitive skills based on the ecological niche within which they live. Building on material we explored in the last three chapters, the feeding ecology of posthuman dogs—whatever it looks like—will strongly influence not only the shape of dogs' bodies and skulls and their social behavior but also will influence the evolution of their minds. Given that dogs around the world rely on humans for food, our disappearance is going to have profound ripple effects on the evolutionary trajectory of dog cognition.

For most current dogs, their ecological niche involves close interaction and interdependence with humans, whether they're regularly homed or living at the edges of human settlements. One of the most exciting speculations about the future of posthuman dogs is how the evolutionary trajectory of their cognitive and emotional lives will shift in response to the sudden absence of what is perhaps the most significant ecological variable in their lives: us. The loss of humans will initiate a tectonic shift in the ecological contexts within which dogs live and a radical shift in the kinds of things dogs need to think about, respond to, and feel. The challenges of adapting to human homes and cities will be replaced by the challenges of living on their own.

Groups of posthuman dogs will inhabit unique ecological niches and their cognitive capacities will continue to evolve in response to the particular difficulties they face. Some of these challenges will be physical (heat, exposure to different elements, altitude, and type of food source), and some will be cognitive. If dogs rely on large prey to survive, cooperative hunting and group living are more likely to evolve within that ecological setting. Cooperative hunting takes a certain kind of brain. Likewise, group living requires certain kinds and levels of social and emotional intelligence. Dogs living in a forested area might develop a more nuanced vocal repertoire than those living in an open desert ecosystem, while dogs in a more open ecosystem might, in turn, develop more nuanced visual signaling. These possibilities are intriguing.

EMOTIONAL INTELLIGENCE

Emotions are "affective" (mood or feeling) responses to stimuli that result in physiological and behavioral reactions. Emotions have evolved because they are adaptive; they regulate and guide behavior. We can go a long way toward understanding the emotional experiences of dogs by looking inward, since dogs experience many of the same basic emotions—or affective experiences—as humans: happiness, fear, anger, disgust, joy, excitement, affection, jealousy, and distress. All dogs experience these emotions. However, as with morphological and cognitive traits, emotional states vary among individuals and there are differences in what individuals feel under what circumstances, the depth of these feelings, and what they do with these feelings. Animal emotions have evolved over millions of years, and it is unlikely that the basic emotional profile of dogs will change very much, unless we consider very long-range evolutionary time frames. Any changes that might take place will be very gradual.

Emotional intelligence is the capacity to effectively recognize and understand one's own emotions and the emotions of others and to use this information to guide one's behavior. Like intelligence more generally, there are multiple emotional "modalities" and different dogs will excel in different areas. For example, some dogs may have exceptionally good self-regulation, which might help control fear responses and anxiety, and allow for self-possession when challenged by an aggressive or unfriendly dog. Other dogs may be good at gauging the intentions and moods of group members, which might help them cooperate successfully with others and avoid agonistic encounters. Still others may be good at diffusing potentially explosive situations in which they might get injured or which might cause a group to lose its cohesiveness.

There has been a great deal of research into human-directed emotions in dogs, such as empathy, the ability to read and respond to human emotional states, and the oxytocin feedback loop that occurs when humans and dogs gaze into one another's eyes. Many dogs are emotionally attuned to humans and bonded to them, but the extent to which these emotional skills in dogs are dependent upon or uniquely directed to humans is often overstated. Without humans, dog emotions may change, but it's hard to say how much difference the loss of humans will really make.

POSTHUMAN PERSONALITIES

Assertive, bold, shy, extraverted, introverted, risk-taking, curious, confident, timid, cautious, impulsive, measured, reactive. The list of personality traits displayed by individual dogs goes on and on, and everyone who has lived with or had much contact with dogs knows that every dog, just like every person, has a special blend of traits—endearing and annoying—that make them who they are.[13]

Personality results from a complex interplay between genetics and experience and influences how successfully animals navigate their environments. Individual differences in personality factor strongly into many day-to-day activities important for survival. For example, personality influences how an individual makes decisions, how they respond to novelty, and how they react to ambiguous signals from others. Moreover, personality is intimately linked with other aspects of cognition and emotion. For example, an individual's approach to solving problems is shaped by personality traits.

We can take research into dog personality as suggestive and use it to frame some general questions about posthuman dogs. For example, will bolder dogs have an advantage over their more cautious peers because they are more open to taking risks and more willing to explore novel situations? Not necessarily. Although we don't have direct evidence from studies on dogs, Samantha Bremner-Harrison, Paulo Prodohl, and Robert Elwood's assessment of behavioral traits associated with the survival of swift foxes in a reintroduction project showed that the radio-tracked foxes who died within six months of release were those who had been rated by researchers as "bold."[14] A 2019 study by Francesca Santicchia and her colleagues found that bold Eastern gray squirrels were more likely to be infected with endoparasites than their risk-averse conspecifics.[15]

Will dogs who are impatient and easily frustrated be at a disadvantage? Again, not necessarily. Some dogs will give up quickly when a problem is difficult to solve, but recalling our earlier discussion of persistence and perseveration, it may be that frustration and a tendency to give up might not be all bad.

Evolution has favored a potpourri of personality traits. Individual variations in personality traits within a population of dogs will almost certainly be crucial to the robustness of groups and to the survival of dogs as a species. A group or pack of all bold or all shy individuals would likely not do very well, nor would a group of all risk takers or all cautious individuals.

HOW WILL POSTHUMAN DOGS COPE?

Individual coping strategies and responses to stress are linked to personality and are shaped by a complex combination of genetics, individual personality traits, and life experiences. Life for posthuman dogs will certainly be stressful at times. Transition dogs will be bombarded by challenges of procuring food, finding shelter, and negotiating interactions with dogs and other animals, and nearly all these challenges will be new and may also be frightening. Many dogs will not survive and the struggle to gain an edge over other dogs will be intense. Even as dogs transition and pups are born into their new posthuman ecosystems, life will not involve a soft bed, a bowl of kibble, and a scratch on the head from a protective human guardian. With people, hard times bring out the best and the worst. Some individuals are resilient and manage to navigate with a cool head and a "glass half full" optimism, while others crumple under pressure, shut down emotionally when under stress or in pain, and lash out at those who might offer social support. The same is true for dogs, and how well individuals can cope with stress and discomfort will influence how well they adapt and survive.

In the 1950s endocrinologist Hans Seyle proposed what has come to be known as general adaptation syndrome (GAS), the three-stage process—alarm, resistance, and exhaustion—by which living organisms respond to and cope with "noxious agents," what Seyle called "stress."[16] As Seyle noted, stress comes in a variety of forms, from hunger, to the presence of a predator, to illness or injury, to hard exertion of running at full speed. People tend to think of stress as a negative ("distress"), but stress can also be useful. "Eustress," or positive stress, can help motivate an animal to act, can encourage innovation, and can build resilience, among other things.

Current dogs are exposed to a variety of stressors, and how they deal with them can help us make predictions about posthuman dogs. Free-ranging and feral dogs are exposed to stressors resembling those experienced by their wild relatives, including discomfort from heat or cold and competition for food and space. It may be that these "gritty" dogs will have developed emotional strategies for coping with stress, which may give them an edge over pampered pet dogs during the Transition. Yet the life of the homed dog is more stressful than most people realize. For example, exposure to long periods of social isolation, such as when their human companion is at work, is very stressful, as are forms of training that involve punishment. So, homed dogs may be developing resilience and coping skills that remain for the most part unnoticed and untested.

Coping styles are individual physiological and behavioral variations in how animals respond to stress. These responses tend to remain consistent over time. For example, behavioral physiologist

Jaap Koolhaas and his colleagues have differentiated between what they call a proactive coping style and a reactive coping style. Reviewing Koolhaas's research, Vindas and colleagues write, "Behaviourally, proactive animals tend to be bolder, more aggressive, dominant and less flexible to changes in routines. Physiologically, proactive individuals are characterised by lower hypothalamic-pituitary-adrenal (HPA) axis reactivity (i.e., lower post-stress cortisol), as well as lower brain serotonergic and higher dopaminergic activity, while reactive individuals exhibit the opposite behavioural and physiological profile."[17]

It isn't that proactive coping styles are good and reactive coping styles are bad. Both can be adaptive for an individual, and variation among individual coping styles within a group will be adaptive for the group. Research into coping styles and strategies in dogs is very limited and has thus far focused on identifying which dogs are struggling most to adapt to a shelter or kennel environment[18] and how coping styles might influence behavior of trained police dogs in a stressful situation.[19] All we can really say about coping styles in posthuman dogs is that they will definitely factor into individual survival.

THE JOY OF PLAY

Play is one of the primary building blocks for the development and expression of cognitive and emotional skills in dogs. Among other things, play involves empathy, cooperation, trust, fairness, theory of mind, and reading others' intentions and emotions. Imagining the role of play in the lives of posthuman dogs invites us to think about why dogs play, how domestication shaped play behavior, and whether play patterns will change under natural selection.

Dogs play for a variety of reasons, not least of which is the simple joy of having fun. The adaptive functions of play include socialization and the development of social bonds, physical exercise, cognitive training, and what Marc and his colleagues call "training for the unexpected"—developing responsive flexibility that might allow dogs to adapt to unpredictable and changing conditions.[20] Play also may be important for establishing long-term social relationships. Of course, not all play is social. Dogs also engage in solitary play, sometimes with an object, such as a stick or pinecone, or sometimes as self-play, such as tail chasing.

Another function of play is to build trust. Data on play behavior in coyote pups suggest that individuals who don't play fairly—who, for example, "break the rules" of play by biting too hard—find it difficult to find play partners. Because they find it difficult to form strong social bonds, they wind up leaving the group and suffer higher mortality than littermates and siblings who remain. Coyotes pay a high price for playing unfairly. So too will posthuman dogs.

One of the most significant changes for homed Transition dogs will be the loss of a primary play partner. For many homed dogs, playing with their human companion is a key part of the relationship and one of the ways in which humans and dogs maintain and form bonds. Play behavior might have been one of the ways in which dogs evolved to solicit attention from humans. It may also have been a key driver in the formation of human-dog bonds. Without humans in the picture, dogs will have more opportunities to play with one another. Under the right circumstances, it's possible that dogs might play with canid cousins, such as wolves, coyotes, and foxes.[21] Dogs share a "play language" with other canids, and their basic patterns of play are very similar. They all use clear and unambiguous signals to tell others they want to play or to keep a play bout going. Dogs might find also some surprising noncanid play partners, such as horses.[22]

Domestic dogs are unique in how much they play and in the continuation of play behavior beyond puppyhood. Selection pressures on dogs during domestication may explain why so many adult dogs enjoy and engage in play. Although research on play behavior in free-ranging and feral dogs is limited, it seems that most if not all dogs play, especially when they are young. Nevertheless, dogs on their own seem to play less than homed dogs. This may be because free-ranging and feral dogs need to spend more time and energy satisfying basic needs, such as defending space, eating and not getting eaten, avoiding and resolving conflicts, and finding mates. Homed dogs most likely play a lot because they have more leisure time. Following this line of reasoning, we might guess that dogs on their own would play more frequently if finding food and remaining safe were relatively easy. One of the factors determining how much puppies play is what you might call "mothering style"—how much mothers allow or encourage play among their children. Research on Amboseli baboons has shown that some mothers limit the amount or exuberance of play in youngsters when times are tough, presumably to conserve energetic resources.[23] Since play is an essential element of development and socialization in puppies, it is certain that they will engage in some play, but probably less than their current-day peers.

How much and with whom posthuman dogs will play is an open question, but it is not too much of a stretch to say that play will be essential to their survival.

In the last four chapters, we've explored myriad ways in which the lives of posthuman dogs might unfold—from morphology and skull shape, to food type and availability, to sex and reproduction, to the structure of social groups, to how dogs' minds and hearts work. All of these are woven together in a complex tapestry.

We have tried to answer, as best we could, our two main questions. First, what will happen immediately after humans disappear? Will dogs survive the transition to life without humans? And second, what will happen to dogs over time, as they continue their evolutionary journey decoupled from human beings? What will dogs look like and how will they behave after tens, or hundreds, or thousands of years under natural selection?

Here are some key takeaways about the evolutionary trajectory of posthuman dogs:

- The physical shape of dogs' bodies and skulls will change. Maladaptive traits (e.g., brachycephaly) will disappear quickly; all dogs will be mutts who look very much like today's feral dogs: medium sized; reddish/brownish fur of one color; pointy ears; longish snouts; medium-length fur (thinner or thicker depending on habitat).
- Finding food will be the primary challenge for posthuman dogs. There will be an abrupt shift in food availability as human-sourced foods disappear. Many dogs won't survive the transition, but their behavioral flexibility and versatility, and opportunism will work in their favor and will help them adapt to novel challenges. Dogs will eat whatever they can get their paws on, and different feeding strategies will evolve over time, depending on ecological niche, local food availability, and competition with other animals.
- Dogs' mating and reproductive strategies won't shift as much as their feeding ecology, but there may be some changes, such as reverting back to one heat cycle a year, more prolonged and ritualized flirting, and greater involvement of fathers and mothers in the rearing of young.
- Dogs will need to sharpen their social skills, including communicating intentions and resolving conflicts. Many

different forms of social organization would work, including the formation of bonded pairs, small groups, and packs. Skills formed during the early socialization period will be critical to survival.

- Behavioral flexibility will be one of the keys to dog success in a posthuman world. Dogs who are best able to adapt quickly to the diverse novel challenges and solve problems will have the best odds of survival. Dogs will have new problems to solve, not only in figuring out how to acquire food but also in navigating, without human help, the complex terrain of interpersonal relationships.

We've provided a *descriptive* account of what might happen to dogs in a posthuman future, exploring which phenotypic traits might be most adaptive and which phenotypic changes might occur at the species and individual levels. We focused on evolutionary questions, with individual and species-level changes being judged as useful or detrimental based on whether they allow dogs to survive and reproduce.

In the final three chapters, we bring the conversation back to the present day and turn attention to some ethical issues raised by the thought experiment. We'll explore what we should be doing now to prepare dogs for a future without us, whether life without humans would be better or worse for dogs, and what light this adventure in speculative biology might shed on the ethical dimensions of sharing a world with dogs, assuming that humans will in fact be around for a long time. In other words, what can posthuman dog futures tell us about how to give our dogs the best possible lives right now?

7

DOOMSDAY PREPPING

The National Geographic television series *Doomsday Preppers* takes us into the minds and lives of various people around the world who believe that catastrophic social or ecological collapse is at our doorstep.[1] They are actively preparing themselves and their loved ones for a bleak future. Some build underground bunkers, others seek out the remotest places on the planet, while still others build high-tech, solar-powered treehouses in the middle of the city. The preppers have packed bug-out bags that are ready to grab and go, have sharpened their knives and bought extra ammo for their guns, have shored up ample supplies of food, have obtained water filters and iodine pills, and have trained their bodies and minds for strength, endurance, and fitness.

Let's imagine the unimaginable: humans are going extinct sometime soon and we need to begin actively preparing our current dogs for a posthuman world. Perhaps we could increase the chances that Transition dogs would be able to survive on their own. For example, there may be skills that we could help them learn, such as acquiring food, chasing prey, finding mates, building alliances, and being alone. Let's continue in the mode of thought experiment, this time with an eye on the present instead of the future. Keep in mind that we aren't advocating any particular form of doomsday prepping, just exploring some of the potential practical implications of the last four chapters.

SURVIVAL 101

If we were to actively begin making our dogs into survivalists, what would our doomsday prepping involve? (This is a separate question from doomsday prepping *for* our dogs, which might involve stockpiling large quantities of dog food.) Feral dogs are already living on their own and are acquiring many of the skills they will need to take care of themselves, as are many free-ranging dogs. As we've stated previously, the most significant change for these dogs during and after the transition is that anthropogenic food sources will dry up. It would be difficult to keep dogs away from garbage and waste, but we could try. Kind-hearted people who deliberately feed stray dogs could gradually stop doing so, although the welfare implications of this might be significant for the dogs, many of whom rely on these food subsidies.

What about homed dogs? Are there things we could and should be doing to help prepare them for our absence? Here are a few areas in which some dog-centric prepping might be useful.

Practical skills: Allow dogs to practice behaviors that might help them survive: let them forage for themselves while out on walks (goose poop, a half-eaten hamburger or tofu hot dog that has fallen out of a trash can), dig holes in the yard, snatch food from the counter or even from our hands; give dogs as much freedom as possible to engage in species-typical behavior.

Physical fitness: Ensure that dogs get plenty of vigorous exercise; try to build endurance (take them on progressively longer walks and then transition them to running with us); build anaerobic fitness by having dogs chase balls until they are exhausted; keep them slim and trim them down if they are overweight; keep their teeth and coats in pristine condition.

Mental fitness: Help dogs develop independence, self-reliance, and self-confidence by providing a stable and predictable environment; let them make choices and exert control over their environment (by installing a dog door); give them communication tools to ask for what they want (to go outside, to play, to eat); expose them to positive levels of stress (e.g., problems that are difficult to solve and involve just the right level of frustration); present them with challenging and enriching work; do less helicopter dog-parenting; allow them to develop the ability to be alone.

Social fitness: Provide appropriate early socialization for puppies to build socially confident, resilient, and competent adults; encourage dogs to form a range of relationships with humans and other dogs; help them develop social skills, including conflict resolution (for example, by not interfering in dog-dog interactions to prevent the rare fight); help them develop excellent dog-to-dog communication skills by fostering healthy relationships and by providing many opportunities for play.

Training and skills acquisition: Refocus training away from human-centered skills, such as balancing a treat on the nose or spinning for food, toward survival-relevant skills, such as focus and impulse control; stop trying to "un-teach" or constrain natural behaviors (marking outside the home, sniffing butts or crotches of people and other dogs, digging, and roaming).

Sports and games: Provide opportunities for nose work, agility, flyball, frisbee, and other dog sports that will help build useful physical skills and mental acuity. Even sports such as pulling sleds or skijoring (pulling a person on skis), which may not build survival-specific skills, would build physical strength and endurance and would be good forms of prepping. We should give dogs ample opportunities to play, particularly with other dogs, because play builds social and communication skills, provides aerobic and anaerobic exercise, and trains dogs to deal with unexpected situations.

SUPERDOGS

Let's say we decide that the best way to prepare dogs for posthuman survival is to ramp up controlled breeding, selecting only for traits that we believe, based on the best scientific evidence, will be advantageous to future dogs. The goal would be to breed a Superdog built to survive without human help (see box 7.1). To do this, we would interfere as much as possible in the reproductive efforts of dogs. Our canine eugenics project would be tricky. It would involve a lot of guesswork and forecasting, since the adaptive or maladaptive value of individual traits depends so much on context. But let's say we decide this is the best way to Doomsday prep dogs. We could start by identifying a spectrum of phenotypic traits, ranging from those we can confidently predict would be maladaptive to those we can confidently predict would be useful, and start breeding with these in mind (see figure 7.1). If we were trying to control breeding as tightly as possible, we would want to surgically or hormonally desex many

Box 7.1. Prototype Dog of the Future

Imagine a fantastical new technology that will print a 3D version (or versions) of a prototype dog. To help design our prototype, our computer is equipped with a post-human, dog-version SIMS ("simulated life") game, where we can input a bunch of variables, including behavioral and morphological traits of dogs, as well as ecosystem characteristics, such as climate, plant and animal communities, and altitude. We can follow a sped-up evolutionary trajectory and see which combination of variables leads to which survival outcomes for dogs. Are there inputs (such as a brachycephalic skull shape) that result in nonsurvival? Inputs, such as behavioral flexibility and generalist diet, that consistently lead to success?

(This is our friend Poppy. She looks a lot like our prototype dog of the future. Photograph by Sage Madden.)

Stage 1: Human Selection for Survival

Less Adaptive Traits (try to select against) ——— Gray Zone ——— Most Adaptive Traits (try to select for)

Stage 2: Natural Selection Takes Over (Stabilizing selection)

Less Adaptive Traits — Gray Zone — Most Adaptive Traits — Gray Zone — Less Adaptive Traits

FIGURE 7.1. Selection for Survival.

of the world's dogs, only allowing reproduction to occur under our direct control.

What phenotypic profile would our Superdog have? We make some predictions based on what we discussed in chapters 3 through 6. We should help dogs become fast runners, agile jumpers, and strategic hunters. We should try to breed for robust health and for dogs who are smart, mentally tough, and resilient. We will want to avoid breeding dogs who are either extremely large or extremely small, but rather aim for dogs in the range of 30 to 60 pounds. The color of the Superdog should perhaps be reddish, like dingoes, or some variation of gray that might help them blend in with their surroundings. We may need to select for morphological and behavioral traits that are quite different from what dog fanciers and image-conscious dog owners are used to. Indeed, Superdogs might make "unattractive" and very poor pets.

There are some problems with this Superdog approach. First, it is hard to know what kind of future world dogs will face and which kinds of traits might thus be most adaptive. Moreover, there is no single adaptive strategy, since not all dogs will inhabit the same ecological niche. The morphological traits that might help dogs in hot, arid regions with sparse vegetation and small-sized prey will be different, at least in certain respects, from the traits that might help dogs in a mountainous ecosystem or in a rain forest. Variation within populations is itself adaptive, and so selecting for one trait or set of traits that we think would be good for individual dogs might end up being maladaptive for dogs as a species.

The adaptive value of cognitive traits will also vary. Dogs who are living in areas densely populated with other dogs may have greater need for nuanced communication and social intelligence than dogs who are living in areas where there will be less opportunity for interaction. However, they may use these skills in encounters with other animals.

Another problem is that evolution is devilishly complex. The idea that we could select for a menu of potentially adaptive traits and have these traits neatly delivered by the Gene Genies is totally unrealistic. As we discussed in chapter 2, when you select for certain traits there are also hitchhiker traits that introduce changes that aren't always anticipated. During the process of domesticating dogs, humans selected primarily for tameness and perhaps also for sociability and trainability. Alongside tameness, many other traits evolved, such as big round eyes, floppy ears, and spotted coats. Even if we selected for a certain smorgasbord of Superdog traits, many unanticipated and unpredictable genetic changes might come along for the ride. Who knows what we would actually wind up with.

REMOVING MALWARE

An alternative approach would be to forget about trying to build a Superdog and instead simply try to remove "malware" from the canine gene pool. Indeed, it may be easier to figure out which traits will be maladaptive than which will be adaptive because certain maladaptive traits won't do well under any conditions and we already know which traits these are.

Certain breeds of dog will be on the "no go" list because they have physical deformities that will make survival extremely difficult if not impossible. Some obvious candidates come to mind. For example, there is no posthuman climate or habitat in which bulldogs will survive. The puppies' heads are disproportionately large and the birth canal too narrow. The inability to breathe and the tendency toward obstructive airway disease certainly won't help matters.

Brachycephalic breeds in general may be at a serious disadvantage, particularly where skull foreshortening is extreme, as it is in pugs and boxers. But dogs with shorter-than-average noses may do just fine, as will dogs with longer-than-average noses. Other malware that humans have deliberately introduced into populations of dogs includes dropped hips, excessively long fur, and stubby legs.

We'll also want to reduce the health problems that have been indirectly caused by years of breeding for physical appearance rather than physical well-being. The best way to address these problems is to allow more genetic diversity into the gene pool, relaxing or getting rid of the notion of breed "conformity." By addressing inbreeding, we might be able to reduce some health problems common to certain breeds, such as hip dysplasia in German shepherds, gastric torsion in Great Danes, skin diseases in sharpeis, and cancers in golden retrievers. Reducing inbreeding and moving away from "breeds" might also reduce the number of dogs who are mentally unhinged. Labradoodles come to mind as a breed that seems to have more than its share of crazies. (Wally Conron, who developed the Labradoodle, said most of the dogs "are either crazy or have a hereditary problem" and says he regrets creating the breed.[2])

One additional step we could take that will be relevant to dogs of the present and near future is to stop surgical disfigurements, particularly cutting off tails and removing ear cartilage. These disfigurements would put Transition dogs at a disadvantage because they reduce dogs' ability to communicate, especially with one another. Tail position, for example, is indicative of mood and can signal aggression, submission, fear, a desire to play, or other emotional states. Ears, too, are an important part of how dogs communicate and variations in ear position signal intentions and feelings. Dogs whose ears have been cropped are missing part of the musculature of the ear, and this may very well reduce their effectiveness when telling other dogs how they feel. Composite signals made up of different combinations of tail and ear position may also lose some of their nuance in dogs who have been surgically manipulated.

HYBRIDIZING

A third potential strategy for prepping dogs through highly controlled selective breeding would be to aim for maximum genetic diversity, building on the principle of hybrid vigor—that a mixing of genes can increase the biological fitness of an organism. We would try to produce the most mixed-breed dogs possible. We're not talking about a Universal Mutt, since no single phenotypic profile is going to fit the bill, but a vast and psychedelic collection of Universal Mutts.

We would still be engaged in hard-core artificial selection, but the goal would be to make each dog a genetic potpourri, as unlike other dogs as possible. To maximize genetic mixing, we could perform a genetic analysis of each dog prior to breeding. Dogs would only be allowed to breed with dogs who are genetically different; if you found Irish setter genes in a male, then you would make sure he doesn't have sexy contact with females of Irish setter ancestry. Stud books and registries could be thrown away, or they could be repurposed to help breeders ensure maximum genetic diversity. There would be no more purebred dogs and the notion of "breed" as we know it today would become obsolete. There would certainly be no more Westminster Dog Shows, unless they could be reconfigured as a sort of Mad Max–style survival games, where dogs had to prove their mettle in finding cached food, heeding potential threats, communicating with other dogs, or speed-digging holes.

Not all outcrossing (introducing genetic material into a breeding line) results in vigorously healthy offspring. Sometimes traits inherited from parents are incompatible and fitness (in terms of reproductive success) is reduced. So, hybridization might be bad for certain individual dogs, but overall species impacts would still likely be positive. Hybridization would diversify the gene pool and go some way toward reversing the ill-effects of the excessive inbreeding of dogs that has occurred over the past several hundred years.

If we are going to go the route of further tightening our grip on artificial selection, we could deliberately interbreed dogs with wolves, coyotes, jackals, and dingoes, taking the idea of heterosis, or hybrid vigor, to a whole different level. Deliberate crossbreeding of dogs with wolves, coyotes, jackals, and dingoes in captivity already occurs, despite welfare concerns and scientific disapproval.[3] Instead of criticizing these renegade breeders, perhaps we should actively encourage them by removing public stigma. We could even take hybridization into the lab and use artificial insemination techniques, using existing captive populations of coyotes, wolves, jackals, and dingoes held in zoos. (Reminder: We are not advocating for any these practices, which we consider highly unethical. Just exploring ideas.)

We could also try to encourage, or at least not discourage, cross-breeding that might occur without our direct intervention. By taking down fences and installing dog doors, we could allow homed dogs to roam more freely. This has some significant downsides, however. Allowing dogs to roam in human-populated areas carries significant risk from car strikes and human predation. Another downside is increased potential for agonistic encounters with other animals that could lead to injury or death for dogs. A final problem is that if we allowed dogs such expansive freedom they might become less inclined to enjoy our company or to rely on us for food and other entertainments. Dogs might gradually feralize, and we might need to rethink how humans keep and interact with dogs.

ZERO GROWTH

In the last section we explored the possibility that current humans should actively intervene in evolution and ramp up control over canine reproduction to create a population of Superdogs who have the best possible chances of survival in a posthuman world. An alternative approach is, again, to aggressively manipulate the reproductive life of current dogs, but this time with the goal of reducing the number of dogs who will face an uncertain and hostile future. The objective would be zero new births, zero growth, and ultimately, zero dogs.

Zero growth would rely on changes to how current and near-future humans acquire dogs whom they want to keep as homed companions. All commercial and hobby breeding would need to stop immediately. A focus on reducing the overall global population of dogs would create stronger incentives for people to rescue existing dogs. If shelters, humane societies, and rescue organizations were the only source for adoptable dogs, and if demand for "new" dogs produced by breeders and sold in pet stores or on the internet dried up, we could squeeze the supply chain shut and the existing population of homed dogs would dwindle over about fifteen years, assuming no new reproduction occurs.

Spay/neuter campaigns are already popular in some countries and may seem, at first glance, like the most effective way to prevent unwanted (by humans) breeding of dogs. Yet there is little evidence that widespread spay/neuter helps control overpopulation of dogs. Many countries where spay and neuter are strongly discouraged, and in some cases even illegal unless necessary for medical reasons, have modest and very well-controlled populations of dogs, and much lower rates of dog homelessness and abandonment. In the US, on the other hand, where veterinary and shelter advocacy groups have been aggressively pushing spay/neuter since the 1970s, dog overpopulation continues to be a vexing issue and many dogs are killed each year because they are "homeless." The problem in the US isn't too many intact dogs having too much sex without human permission. The problem is us: too many breeders are trying to make a buck by selling puppies. Indeed, at least two-thirds of all litters of pups in the US are not "accidental" but have been intentionally bred by humans.[4] Purposely creating new puppies—often by means that cause a great deal of suffering to mother dogs—while healthy dogs languish in shelters simply makes no sense.

It is hard to say precisely how the shifting population dynamics of homed dogs would influence the population dynamics of free-ranging and feral dogs, but an overall reduction in pet- keeping and numbers of homed dogs would probably also lead to a constriction of overall global dog populations. Intervention in breeding of nonhomed dogs would require some plan for catch-neuter-return, or perhaps the development of a chemical sterilant that could be delivered by way of baited food.

The zero-growth approach builds on the assumption that dogs face an ugly future with very high mortality and that we should try to prevent suffering by no longer bringing new dogs into the world. We should give our best care and attention to dogs who are here now, and revel in these perhaps final years of sharing our lives with canine companions as the human-dog bond reaches its denouement. This response is focused mainly on the prevention of suffering for individual dogs.

HANDS OFF

Throughout the process of domesticating dogs, humans have manipulated and exploited the life history traits of dogs. We have decided which traits will survive and be perpetuated, imposing human-chosen life-history strategies onto dogs. Natural selection wouldn't, for example, produce dogs who can't give birth naturally or who can't breathe when they run. So far, we've talked about trying to prepare dogs for a posthuman future by intensifying human-directed selection and shifting the priorities away from human aesthetic preferences and toward canine survival. We keep our hands on the ship's wheel, tighten our grip, and shift the direction of travel. This approach assumes that we have ample foresight and understanding to shape the evolution of dogs in ways that will be beneficial to them in a future without us. But we may not. It may be that Nature knows best and we should hand the wheel to her. We could sit back and let natural selection take over now: stop neutering and spaying, stop trying to prevent random breeding, stop interfering in dogs' mate choices, and give dogs as much freedom of movement as possible so they can mingle with others of their kind. We could still live with dogs as companions, although over time they might become less tame, less docile, and less interested in being our pets. More and more dogs would become free-ranging and then feral. The giant cruise ship of dog domestication would drift forward for a time, slowly discharging 20,000-odd years of built-up momentum, and then gradually start to follow a new trajectory.

We have several different strategies that could be used to help tip the scales in favor of survival for posthuman dogs: strong human-directed selection for adaptive traits, prevention of breeding to reduce the number of future dogs, and letting natural selection take over now so the process of re-wilding can get underway as soon as possible. Which of these approaches is the best? It's hard to say. We like a combination approach, hands off, mixed in with a little bit of coding for success and removing malware, and heterosis. This combination approach seems like it will provide dogs the best chances.

So far, we've talked about how to prepare dogs for the future and how to shape our interactions with them. Let's now briefly discuss one other potential response to a posthuman future, one that is chilling even to contemplate and which we ourselves find abhorrent: the widespread killing, for "compassionate reasons," of as many dogs as we can.

PROPHYLACTIC KILLING

Anxiety about the future, particularly at this critical juncture when climate-related catastrophes and pandemics are real and growing threats, is something that keeps many people up at night. As we are completing this book, the COVID-19 pandemic is having major impacts on human activities and is having a profound ripple effect among the world's dogs.[5] For those living with a dog, these challenging times are bound to involve some concern for their canine companions, as they should. And it is worth asking, if only to shudder at the possibility: If we knew that humans were on the way out, would "prophylactic killing" of dogs be an ethical response?[6] If you knew that a nuclear bomb were headed your way, would you—if you had the means to do this without pain or fear—take the life of your dog to prevent him or her from suffering?

We have a disturbing historical example of this extreme response to impending disaster. As Hilda Kean describes in *The Great Dog and Cat Massacre: The Real Story of World War Two's Unknown Tragedy*, prior to and in preparation for the bombings that the English thought were inevitable, people were asked to bring their dogs and cats into the "pound" so they could be killed. Hundreds of thousands of family pets were preemptively euthanized. It was considered a kindness to protect these animals from the potential terror of the bombings and the potential loss of their human owners.[7]

Our overriding conclusion is that prophylactic killing is a very, very bad idea. Humans cannot forecast the future and even catastrophic scenarios in which there is a large-scale and immediate threat not only to humans but to other beings are unpredictable. Individual dogs should be given the chance to survive. We need to be open-minded about the possibilities for canine happiness—including the happiness of our own special dogs—in a world without us.

Throughout this book we've made the case that many dogs will survive without us and will do quite well. We've challenged the assumption that dogs must have people—and must have a specific person or family who lays claim and "owns" them—to want to continue living. Even intensively homed companion dogs who love and are bonded to their humans have a good shot at surviving and thriving on their own. Feral and free-ranging dogs who are already living partly or mostly on their own, and who not only scrape by but also often flourish, provide the strongest counterargument to the notion that dogs couldn't survive without us.

WHAT IF DOOMSDAY DOESN'T COME?

Let's suppose that we have made a concerted effort to prepare dogs for a posthuman future but our forecasting was inaccurate. Doomsday doesn't come, and humans are still here. Although many aspects of prepping we discussed are morally problematic, there is nevertheless some significant alignment between doomsday prepping and efforts to provide current dogs the best possible life in a world *with us*. Thinking through the various possible forms of prepping—particularly those that seem most ethically offensive—can perhaps help illuminate problems and possibilities in current human-dog relationships.

Would the physical traits that benefit dogs in a future without us be the same as the traits that would benefit dogs if we're still around? To a large degree, yes. Regardless of what happens in the future, we should make every effort to allow dogs a physically robust and healthy life. Overall, breeding for survival traits would produce a population of dogs who are physically and mentally fit and who would be behaviorally versatile enough to adapt to a range of living situations. It would also go a long way toward reversing the damage caused by decades of breeding practices aimed at making dogs into pedigreed ornaments, into canids trapped inside the wrong kinds of bodies. If we deliberately control breeding by dogs, our focus should be the physical well-being of individual dogs, not their appearance. We should reduce inbreeding, eliminate traits that contribute to poor quality of life or shortened lifespan, increase diversity in the gene pool of dogs, and move beyond the obsession with pedigree and breed.

Another aspect of prepping that aligns with giving dogs the best possible life is the idea of physical form being suited to ecological context. As it is, humans acquire dogs with little attention to the suitability of the dog to local habitat. Sure, humans help dogs "overcome" environmental unsuitability by keeping them indoors or buying them puffy jackets and boots. But even with human interventions, the mismatch between dogs and their habitat compromises their quality of life. It isn't that huskies can't live happy lives in Palm Springs—they can. But they struggle when it is 100 degrees, and their quality of life is impacted because they are restricted in when and how much time they can spend outside on walks, playing with other dogs, and so forth. Why, if we have a choice, wouldn't we align the needs of a dog with the local conditions within which they will live their lives? There are many other ways in which humans might do better in aligning the needs of a dog companion with the opportunities and limitations of our own lives. High-energy dogs do best with high energy people, and so forth.

Dogs who are ready for a posthuman future may not make ideal companion dogs, at least by current-day standards. But perhaps it isn't dogs who need to change as much as it is our perceptions about and expectations of dogs. A good many people who bring dogs into their homes seem surprised and even offended when their dog behaves like a dog. People want a dog who doesn't have fur, or who doesn't bark, or whose paws never get muddy or scratch the new wood floor. Many species-typical behaviors that dogs are highly motivated to perform are labeled as "naughty" by people and even by veterinarians. In a posthuman world, dogs will need to *be dogs*. They will need to perform the suite of behaviors for which they have evolved: barking, digging, sniffing butts, chasing squirrels, rolling in dead stuff, running fast, playing with other dogs. These are behaviors we should allow, even encourage, in our dogs now.

This leads us now to a final question, which we pose in the next chapter: Are the compromises that we ask of dogs, in living with us, greater than they should bear? To put it more bluntly, do dogs have more to gain from our disappearance than they have to lose?

8

WOULD DOGS BE BETTER OFF WITHOUT US?

Would dogs be better off without us? This may be a difficult question to consider if you live with a dog, love dogs, and find beauty in the enduring loyalty of the human-dog partnership. If you are reading this book with a dog curled up next to you on the couch or on her fluffy dog bed happily licking peanut butter out of a Kong, this question might even be too painful to contemplate: How would *my* dog survive, naked and afraid, set adrift in a frightening new reality, without me to keep her safe? Yet try to imagine for a few moments not only what your dog might lose, but what she might gain. Better yet, think about the whole range of individual dogs who currently share the planet with humans and consider the potential losses and the potential gains of having the world to themselves. And think about dogs who might come after the Transition, who have never known life with humans. Maybe dogs as a species would have a better go of things on a planet that they didn't have to share with people, if the 20,000-year-long domestication experiment—which, arguably, has had its problems—were called off once and for all.

Dogs would be challenged by living on their own in a posthuman world. But a posthuman world is also full of what you might call "dog possibilities"—the various ways in which dogs would adapt, innovate, and expand their experiential worlds. We've seen that there is far more to the lives of dogs than being a house pet, spending the day chasing balls, barking at the postal delivery person, or waiting anxiously for their Person to come home from work. A Dogs' World is a bustling place, with dogs working on their own and with others to solve the puzzle of survival and to reap the rewards of life. Trying to catalog what dogs might stand to gain and lose if humans disappeared can help bring into focus some of the ways in which humans make life hard for dogs. More pertinent for those of us who live with companion dogs are the potential insights about how we might, without even realizing it, be asking our dogs to live in ways that constrain who they are and who they might become, the many ways in which we compromise the "Dogness" of dogs. Having a sense of the whole experiential range of dog possibilities may help us become better companions to our dogs.

To explore whether the dog sitting next to us on the couch is fantasizing about a humanless world, we've tried to identify the potential gains and losses for dogs in a world without us. As you might expect, the question "Would dogs be better off?" does not yield a simple "yes" or "no" answer, and the further you dive into the question the murkier the waters become.

VARIABLES AT PLAY IN THE GAINS AND LOSSES GAME

We've constructed a comprehensive list of what dogs stand to gain or lose if humans go extinct (see tables 8.1 and 8.2), which we'll come to in a few moments. First, though, here are a few thoughts on why judgments about gains and losses are complicated.

What dogs may gain or lose as a species is distinct from what an individual dog stands to gain or lose. The sudden disappearance of humans will result in broadscale losses at the individual level. Many dogs will be ill-equipped to survive, not having had any lived experiences of obtaining their own food, finding shelter, or forming a workable pair bond. Depending on how humans disappear, individual dogs in captivity—for example, inside homes with no way to get out, or locked inside shelters or laboratory cages—will perish. Overcrowding of dogs in some areas may lead to intense competition for scarce food resources. Moreover, large numbers of individual Transition dogs will be unable to reproduce because they have been desexed, so even if individuals manage to survive, they will be at a genetic dead end. Nevertheless, enough dogs may survive this first wave that viable populations will take hold in habitable ecosystems. Dogs as a species may very well go on to flourish.

The gains and losses for Transition dogs will be unique and will depend a great deal on where a dog begins this unprecedented journey into a posthuman future. The unique characteristics of where and how each Transition dog is living when humans disappear will greatly influence what challenges they face and what is experienced as a loss or gain. How well they cope will depend

Table 8.1. When we're gone . . . what do dogs stand to gain or lose?

	Physical	*Social*	*Psychological*
Gains	Freedom of physical movement No human constraints, like collars, leashes, fences, cages No intensive captivity, such as in puppy mills, laboratories, or dog meat farming No experimentation No forced breeding No abuse, sexual exploitation, dog fighting No killing of healthy dogs No artificial selection for maladaptive traits (brachycephaly) Lower levels of obesity Potentially better nutrition Greater range of sensory experiences (e.g., can more fully use olfactory sense) Natural level of hormones and development Physical activity budgets would be chosen by dogs themselves, not by humans No desexing (hormonal implications; susceptibility to certain diseases; growth plates/development; etc.) No surgical mutilations, such as tail docking, debarking, and ear cropping No shelters and no shelter-related mortality Reduction in breed-specific genetic disorders	Ability to act independently and to make free choices More control over choices Freedom to choose friends Freedom to mate with whom they choose and when they choose Freedom to engage in parental and alloparental behaviors Freedom to engage with littermates and siblings Freedom to be alone or have downtime Freedom to form groups and engage in pack/group behavior	No psychological sequelae from captivity No fear of or stress from human punishments, violence, or confinement No human-induced fear of or stress from unpredictability and inconsistency No human-induced trauma Possibilities for having companionship with dogs and other animals No human-induced learned helplessness Greater sensory stimulation Freedom from sensory deprivation Freedom from sensory overload in human environments Heightened sense of satisfaction (e.g., engaging in work) Agency Choice Lower rates of anxiety and depression Less likelihood of experiencing boredom

Table 8.1. (continued)

	Physical	*Social*	*Psychological*
Losses	No veterinary care No pain management (medicines, massage, acupuncture, palliative care, pain medications, etc.) Potential exposure to diseases Loss of physical comfort No regular meals Potential for nutritional deficiencies Greater exposure to predation Greater exposure to the elements No human-provided safe zones No anthropogenic food sources No human-provided control of parasites No human-provided hygiene Loss of potential health gains from spay/neuter No euthanasia to relieve suffering or pain	Loss of human social companionship Humans won't plan for and facilitate social interactions. Less arranged "free time" to play with friends Humans won't help mediate conflicts.	Greater fear of predation Greater fear of ecological unpredictability No access to mental health interventions such as anxiolytics

on an individual dog's personality, past experiences, learning, social and emotional intelligence, and physical attributes.

Dogs currently live in wildly diverse relationships with humans, and while some dogs may keenly miss humans, others will be glad to see us go. A homed dog with a well-informed, motivated, and empathic human caregiver has more to lose than a dog caged at a research laboratory or in a puppy mill. Feral dogs will miss the enormous piles of garbage that humans produce but may not experience any loss of human companionship. Although the challenges for homed, free-ranging, and feral dogs will be different, the loss of humans and the transition from human selection to natural

Table 8.2. Gender differences in gains and losses?

	Males	*Females*
Gains	More sex Retention of genitals Normal levels of testosterone Opportunities for fatherhood Sexual pleasure Choice of mates/sexual partners	Reproductive freedom/control Retention of ovaries Normal levels of estrogen and progesterone Opportunity for motherhood Opportunity to raise young for the full term of motherhood Sexual pleasure Choice of mates/sexual partners
Losses	Competition for mating privileges Coitus interruptus (having another male break the copulatory tie before insemination occurs)	No veterinary intervention during difficult births

selection will be abrupt, and it won't be pretty for many of the dogs on the planet.

There will be far fewer posthuman dogs inhabiting the planet. A reduction in total numbers should not necessarily be viewed as a loss because arguably there are too many dogs, their population having been bloated by intensive human breeding and careless pet-keeping practices. The size of dog populations, especially in dog-dense areas, will need to be much smaller to be sustainable, with sustainability depending on the carrying capacity—the maximum population size of a species that can be sustained within a given environment—of different habitats in which dogs are trying to survive.

Posthuman dogs may form short- or long-term groups. What might be a gain for a group isn't necessarily a gain for all individuals within the group, and much will depend on who else is in the group and the ecological conditions with which the group must contend. As we've noted, groups of animals tend to be most robust when they contain a broad range of behavioral phenotypes. It may be good for a group to have a combination of high-ranking and low-ranking individuals, but life might be difficult for those of lower rank.

If humans disappeared, some gains and losses would be felt immediately, such as loss of human food subsidies and the gain of freedom from physical constraint, but the effects of human disappearance will reverberate and shift over generations.

GAINS AND LOSSES

Our list of gains and losses is divided into three categories: physical, social, and psychological. There will be trade-offs among these variables just as there are among life-history traits. You may think of things to add to the lists or better ways to organize these speculations. You may bicker with us about what we've put in the "Gains" column and in the "Losses" column, but this is a good beginning for further conversations about posthuman dogs and, more importantly, for current dogs and the people who care for and about them.

We'll examine a few examples from the table of gains and losses and provide a bit more detail to give a sense of how complex and sometimes counterintuitive they can be. We won't try to cover every single entry in detail. Some are obvious and don't need elaboration; for others there isn't much to say.

PHYSICAL GAINS AND LOSSES

Dogs will lose some of the key perks of being our companions: a steady supply of nutritious food, fresh clean water, soft bedding, and shelter from the elements. They will lose a range of beneficial veterinary care, such as vaccinations, disease and pain management, wound care and antibiotics, grooming, and parasite control.

These losses would be countered by some significant gains. As we noted in chapter 4, humans are the main cause of mortality in dogs, through a combination of deliberate extermination of dogs in parts of the world where rabies is a serious problem, along with car accidents and the "kind-hearted" killing of dogs in parts of the world where dogs without human homes are considered "homeless" and are not allowed to live on their own without a human "owner."

Dogs would no longer be prey to human cruelty and exploitation. They would no longer be tortured in laboratory research facilities. Female dogs would no longer be forced into the role of breeding machines. Dogs also would be free from human sexual abuse, free from participation in sports such as fighting and racing, and free from the extreme forms of physical abuse often inflicted on dogs by humans. Dogs would no longer be raised and sold as meat. The various inflictions of cruelty by humans are not rare, isolated instances of misuse. These are experienced by millions of dogs every day.

Beyond the obvious cruelties inflicted on dogs by humans are the underacknowledged sufferings of homed dogs whose lives are experientially uninteresting, who have few opportunities to engage in species-normal behaviors, and who experience chronic low-grade stress. Roughly 80 percent of dog owners in the US report that their dog has "behavioral problems." Among the behaviors that are commonly reported as "problematic" by owners, at least half are actually normal dog behaviors: digging, attention seeking, running off or escaping from the yard, stealing food, barking, pulling on the leash. Dogs are not allowed to be themselves and are often punished just for being dogs. Unsurprisingly, millions of dogs suffer from psychological problems such as anxiety and depression.

Without humans, dogs would be freed from the constraints of being square pegs forced into round holes, of being dogs expected to live and act like furry people. They would also be freed from the many quotidian constraints imposed by humans, such as leashes, crates, fences, and shock collars. Time and activity budgets would be chosen by dogs themselves. Dogs would be free to engage in a full range of natural behaviors and would gain what we might anthropomorphically call "self-determination."

Many of these physical gains and losses will be interconnected with social and psychological gains and losses. For example, the loss of physical shelter may be linked to increased experience of fear and discomfort. Having self-determination in the physical realm would also benefit dogs in the psychological realm: control over decision-making links with feelings of satisfaction, self-confidence, and happiness.

SOCIAL GAINS AND LOSSES

Dogs have a lot more to gain than lose in the "social" category. They'll suddenly be free to interact with other dogs, something that is difficult for most homed dogs and which may also be constrained in free-ranging dogs. Dogs will be able to make friends, form alliances, and engage their full repertoire of communicative and other social behaviors. And after the Transition period, all dogs will be reproductively intact and will be able to experience a range of sexual and parental behaviors. Because humans may select for certain behavioral phenotypes—for example, bold, friendly, outgoing—when humans aren't around, a broad spectrum of behavioral profiles will be expressed among populations of dogs.

Still, a posthuman world won't be like one great big dog park. Dogs may have the opportunity for social engagement, but not all social interactions will be friendly. Agonistic encounters with dogs and other animals will be part and parcel of social freedom, and these can lead to injury and sometimes death. Even injuries that don't appear to be serious can render an individual reproductively inactive. Dogs will no longer have humans around to mediate conflicts or facilitate social interactions between and among dogs and between dogs and other animals.. Humans will not provide help for socially awkward dogs who wouldn't survive well on their own. Dogs who are socialized by their canine mothers, fathers, and siblings—and not by well-meaning but dog-illiterate human "parents"—will be better prepared to interact with other dogs, which will be a net gain.

Perhaps the most significant social loss for dogs will be the dissolution of the bond that has evolved throughout the process of their coevolution with humans. Domestication has shaped the social behavior of dogs in multiple ways—think of the range of human-directed behaviors in dogs, such as the oxytocin-feedback loop (a positive feedback loop in which the release of oxytocin stimulates actions that release still more oxytocin, so that everybody feels the love), gaze-sharing and gaze-following, and the exquisite attunement of dogs to human emotional cues, such as facial expressions. Dogs are evolutionarily heavily invested in humans. Can this substantial investment be cashed out by dogs into their relationships with one another?

PSYCHOLOGICAL GAINS AND LOSSES

Humans now strongly influence the psychological state of many, perhaps most dogs around the world, both for better and for worse. We protect dogs from some of the stresses and fears they might experience if they were fending for themselves, such as fear of predators and the stress of unpredictable meals and environments. In many instances we also provide the comfort of having a trusted companion. Yet humans also terrorize dogs: feral and free-ranging dogs are often hunted down and killed. We impose profound psychological suffering on dogs who are held in puppy mills, research laboratories, dog-fighting operations, and other intensively captive settings. Even homed dogs are confronted with a range of psychological challenges, such as our unpredictable comings and goings, our overpressuring them to perform, our tendency to miscommunicate, our frequent and often confusing punishments, and our unhealthy emotional codependency.[1]

Some aspects of the domestic relationship may not even enter our minds as cruelties because they are such familiar and unquestioned elements in our human-dog partnership. One such act is the practice of taking puppies from their mothers and making them our own. The process of "getting a dog" may feel joyous from our perspective—we bring a sweet-breathed, whimpering, vulnerable ball of fur into our home and begin forming a bond with our new best friend. Our caretaking instincts are activated, and we feel love brewing. But from the dogs' perspective, the experience may not be so rosy, either for the pup or the mother as we subvert natural canid behaviors such as parental caregiving and attachment. As David Brooks writes in *The Grass Library: Essays*, "We impose so many wounds upon the animals we think of as our companions; taking them away from their mothers so early—barely weaned, if that—is only the beginning."[2]

DYSTOPIA, UTOPIA, OR DOGTOPIA

It seems that dogs might just be better off without us. The list of potential gains is considerably longer than the list of losses. Moreover, as a reader commented on an essay by Jessica, most of the items in the Losses column are replaceable. For example, nutritious food, water, and shelter can be found in nature, friendship can be found in the pack, and toys would not be needed if dogs weren't forced to spend all day at home.[3]

Still, it's impossible to make a blanket statement like "all dogs will be better off when we're gone." It isn't simply a number count of how many gains and losses we can list. Some dogs will gain; others will lose. Some gains and losses are more important than others, and each individual dog will experience gains and losses differently. A profound loss to one dog may not matter all that much to another. And there will be trade-offs. The loss of veterinary care might be balanced by gains in freedoms such as being able to choose with whom to mate, the absence of surgical manipulations, and greater opportunities to make friends. A loss of human-provided food might be balanced against more freedom to choose when, where, and what to eat.

Surviving is not the same as thriving and flourishing. Dogs on their own will survive (some of them, at least), but will they have what they need to thrive? Physical needs must be met first: dogs will need to find enough to eat and good enough places to hide from predators and from the elements, and will need to have enough energy and opportunity to reproduce. If you can't eat, or you become someone else's dinner, there isn't much to talk about. Also, if mere survival is all that will be available, is this a decent existence?

Physical, social, and psychological needs are intertwined, and it is hard to separate them. Freedom of physical movement, for example, is linked to psychological gains of increased agency—the freedom to make choices and act independently—and greater happiness and satisfaction.[4] Ultimately, for a posthuman future to be a place we can bear considering for our canine companions, the social and psychological elements that provide joy, comfort, and excitement will also need to be available.

This discussion brings us to one of the questions that motivated us to write our book in the first place: "What is the 'best possible life' for a dog?" It is relatively easy to come up with scenarios—

both present and future—that seem awful for dogs. It is much harder to try to construct an imaginary world in which dogs have everything or nearly everything they need to be happy. What might a dog utopia—a Dogtopia—look like? And would it include humans? Will dogs always be looking sideways into the copilot seat or can they happily fly solo? For posthuman dogs, the sky is the limit.

9

THE FUTURE OF DOGS AND DOGS OF THE FUTURE

Returning now to the question that motivated our book, whether dogs will survive in a world without humans, we have a whole array of answers. The transition years will be hard because of the loss of human support, especially food subsidies, and more varied and unexpected encounters with dogs and other animals. These new demands will require behavioral, neural, anatomical, and physiological adaptations, and a healthy dose of luck. Natural selection will quickly weed out morphological and behavioral traits that are maladaptive, such as extremely foreshortened snouts and too many skin folds. Dogs with no experience trying to survive on their own will need to adapt quickly to their new situation, and some will handle this better than others. The geographical distribution of dogs will shift and shrink, with populations concentrating in ecosystems and climates where they can best survive without human support. But overall, many dogs will survive and even thrive.

We've provided a range of ideas about what the evolutionary trajectories of posthuman dogs might look like. A recurring theme has been trying to understand and appreciate the ancient impulses and memory traces that still lurk in dogs' brains—the indelible engrams that still influence what they do and how they feel and which will shape how they do without us. But we are still a long way from understanding who dogs might become if human selection were abruptly halted and natural selection took over. There are few straightforward answers—which makes it a perfect subject for a thought experiment in imaginative biology.

THE POWER OF IMAGINATION

Taking an imaginative journey into a world without us provides fertile ground for scientists who are studying dogs now. Although dogs are everywhere around us, we still have a lot to learn about them. We hope to have completely dispelled the idea that dogs are among the most thoroughly unnatural animals on the planet and that only "natural" animals are worthy of study. Dogs are just as natural as wolves and coyotes and other canids, their biology and behavior just as interesting. Dogs should take their rightful place in *Canids of the World* and deserve more than a single paragraph.

That said, it is important to emphasize that dogs present an array of challenges to scientists. Dogs inhabit diverse ecological niches: for the approximately one billion dogs on the planet there are approximately one billion distinct canine ways of life. Dogs employ a wide range of strategies to survive. They have been subjected to both artificial and natural selection, and these evolutionary threads are hopelessly tangled together. Does this make dogs difficult to study? You bet. But scientists manage to study Adélie penguins in Antarctica, profoundly secretive wolverines in the snowiest and most remote parts of Montana, birds who nest on cliff faces hundreds of feet off the ground, and worms at the bottom of the deepest oceans. If researchers are innovative enough to study these species, they surely can find creative ways to study dogs.

What kinds of studies need to be done now? Some of the most important research will investigate life-history traits and trade-offs; how dogs function within diverse ecosystems and niches within them; the natural behavioral repertoire of dogs and how this may change in a world of fewer or no humans; and how dogs might decouple from humans over time.

Life-history studies of *Canis lupus familiaris* can help us get a better handle on topics such as mating strategies, the role of fathers and helpers in raising young, space use, feeding ecology, and relationships between body size and habitat. The best way to understand how dogs are going to do without us and what their lives might be like is to study free-ranging and feral dog populations around the world. Researchers already studying free-ranging and feral dogs—and whose work we have relied on throughout this book—are collecting data that will help us move from speculation to more data-driven hypotheses. This research can add to life-history studies and can go a long way toward busting myths about who free-ranging dogs are and how they live.

Just as myths about free-ranging and feral dogs abound, so too do assumptions about homed dogs, and these often limit research into canine ethology by constricting the type and range of questions that are asked and the data that are collected. One of these assumptions is that the natural environment or ecological niche of dogs is human households. As we have made clear throughout this book, it is hard to say exactly what the natural home or ecological niche of the dog is. Dogs around the globe inhabit diverse ecological niches, only some of which center around a human home—and human homes can be as different from one another as the Sahara is from the Arctic. Moreover, many dogs will wind up inhabiting more than one niche during their lifetime, with the most obvious example being a homed dog who is abandoned by their human and takes up life on the street.

Most studies of dog cognition focus on homed dogs and take place in canine cognition labs or via surveys of human caretakers. These projects produce interesting results about homed dogs whose owners bring them to labs or fill out surveys. And unlike much research conducted on animals in laboratory and wild settings, the protocols inflict no harm on research subjects. In the lab, dogs volunteer their time, seem mainly to enjoy the work, and are never subjected to invasive or painful protocols. The problem arises when these results are generalized to all dogs, or even to all homed dogs. Context is exceedingly important.

Could we say which behaviors are natural for dogs? Are they the behaviors that homed dogs display? Are they the behaviors in which wolves engage? Or are they somewhere in between? "Natural" behaviors for dogs seem harder to pin down than natural behaviors for wolves, because the intervention of humans into dog behavior is so pervasive. Take peeing, for example. For male wolves, lifting a leg to pee is a natural behavior and for female wolves, squatting to pee is a natural behavior. Male dogs also typically lift a leg, while female dogs squat. So far so good. But while it would be natural for a dog to pee in their territory and home range, we don't allow dogs to pee in human houses or on other people's lawns or in certain other locations because we don't like it when they do. We also usually don't allow them to roll in dead stuff, gnaw on a deer leg in our bed, dig holes in the yard, or roam around the neighborhood looking for sexual partners. At least for homed dogs, the suppression of natural behavior begins at birth and extends throughout life. There are things dogs don't do because humans don't let them. There are also things they don't do because they don't have to, such as hunting and defending food. Living in partnership with humans, dogs' natural behaviors have shifted.

It would be fascinating to better understand the mechanisms of dogs' dependence on humans in terms of feeding ecology, mating strategies, and cognitive and emotional capacities. For example, it is often said that all dogs everywhere are, on some level, dependent on human food sources. But there are two ways of describing and understanding this: (a) dogs need anthropogenic food sources; or (b) dogs have successfully taken advantage of anthropogenic food sources. Scientific studies of dogs have tended to frame dependence in the first way. But it may be that the second framing is also accurate, and the second way of framing this dependence leads to a different set of answers about how dogs might do on their own.

THINKING ABOUT THE FUTURE, LIVING IN THE PRESENT

Our journey into a posthuman future is relevant not only to scientists who study dogs, but also to the millions of people who share their home with a canine companion. By thinking about a future when dogs go wild, we can learn a lot about dog-human relationships right now. What are some of the takeaway messages?

1. There is no universal Dog. We need to be careful not to make sweeping generalizations about what dogs do or

don't do, or even what's good or bad for dogs. The focus needs to be on individuals.

2. Certain traits make animals better able to adapt to different conditions and more likely to survive and thrive in a challenging future. It is hard to think of a scenario, future or present, in which maladaptive traits are good for an animal. So, we should stop selective breeding for traits that are maladaptive and only serve humans.
3. Those who live with homed dogs ought to consider allowing their dogs to engage in a wide range of species-typical behaviors. We have a much better handle now than we did at the beginning of the book on how natural behavior might be cashed out for dogs. Humans can be more thoughtful in their approach to living with dogs if we use the best canine science to understand who they are. It is also useful to see dogs as wild animals, as part of a wide variety of ecosystems, including human homes. Dogs are not outside of nature.
4. Dogs have a wide range of natural habitats and live alongside humans in diverse ways. But some habitats inhibit the ability of dogs to be dogs in any meaningful way, such as laboratories, dog meat farms, or puppy mills. Some habitats are less obviously inhibiting, but nonetheless may greatly limit a dog's ability to live an interesting life. Little dogs who are bought as fashion accessories, and who have their nails painted and are taught to "go" inside on fake turf, are not really allowed to behave like dogs. Dogs who are obsessively helicopter-parented by their human guardian can have their ability to engage in normal dog behaviors seriously compromised.
5. Many people love their dogs in ways both rich and diverse. Because dogs tap into such a deep reservoir of human

compassion, they may serve as a gateway species and catalyst for expanding compassion and empathy to other nonhumans and also to humans.

6. It isn't all that pleasant to think of a world in which we're no longer here, but there are many reasons to believe that when we're gone dogs will survive and life will go on. It is healthy for us to begin decentering the human. When we decenter, then real, fruitful nonanthropocentric thinking can evolve.

DOG FUTURES

When humans think about dogs, we tend to look backward. Scientists try to understand who dogs are by decoding their past, by trying to answer the elusive questions of when, where, and how wolves became dogs. They sift through the archaeological record, compare minute differences in shapes of skulls and sagittal crests on fossil bones, and analyze tiny strings of wolf and dog DNA looking for clues.

Scientists aren't the only ones who look backward. Humans who currently live with a dog also tend to see, behind their dog, the large shadow of wolf. One of the most common responses we hear from people when we talk about this book is "Of course, dogs will just go back to being wolves. After all, this is where they came from."

However, evolution doesn't work backward. It unfolds into the future and there is no going back. There was no single, forward-directed trajectory from wolf to dogs. The domestication of dogs occurred piecemeal, in different places and over a span of time. It is impossible to pinpoint when wolves became wolf-dogs and wolf-dogs became dogs. Just as there are multiple origins of domesticated dogs, there also will be multiple origins for dogs becoming less dependent on us. Feral dogs are already further along on this trajectory toward independence than dogs living in city apartments. They are already residents of wild communities. Where dogs will go next is toward something altogether new.

Will the most important event in the continuing evolutionary journey of dogs be the disappearance of *Homo sapiens*? Yes. Is contact with humans a necessary part of what it means to be a dog? No. Although the loss of humans would be catastrophic, dogs would carry on.

Our take-home message centers not solely on an imaginary future, but also on the present. In thinking about who dogs might become without us, we may gain fresh insight into who they are now and how our relationships with them can best benefit us both. Looking at what dogs might gain were humans to disappear can help crystallize some ways in which we compromise their Dogness, and greater awareness of this might help us give dogs greater independence and freedom right now. Dogs' close partnership with humankind has strongly shaped the evolution of the species; they wouldn't be who they are today if it weren't for us. But coevolution, by definition, goes both ways: we would not be who we are today without dogs.

Many different futures are possible for dogs, and some are better than others. But a future without humans isn't as bad as you might, at first, imagine.

People may say to the dog curled up peacefully on the rug by the fireplace, "Whatever would become of you without me?" Perhaps under their breath the dog whispers a soft reply, "A lot!"

ACKNOWLEDGMENTS

Our deepest gratitude to Christie Henry for her support of our work over the course of many years. Thank you to Alison Kalett for seeing potential in our early ideas and helping to bring these ideas to life, to Dana Henricks for her excellent and entertaining copyediting, and to Natalie Baan, Whitney Rauenhorst, Matthew Taylor, Kate Farquhar-Thomson, and everyone else on our team at Princeton University Press. Amron Gravett is an indexer extraordinaire. Marco Adda graciously provided images of free-ranging dogs in Bali and answered many questions about these remarkable animals. We are grateful to Sage Madden for sharing photographs of Poppy, Prototype Dog of the Future. Rick McIntyre, L. David Mech, Douglas Smith, and Robert Wayne clarified what is known about inbreeding in wolves. Brad Smith and Brad Purcell provided valuable insights into Australian dingoes, and Andrew Rowan continually updated us with information about the best estimates of global populations of dogs. We also thank Brooks Fahy, Michael W. Fox, Betty Moss, Paul Paquet, and Michael Worboys for sending us relevant sources, and Jonathon Turnbull and Adam Searle for useful discussions about how to classify dogs and what is really happening at Chernobyl. Thank you to Mark Derr for reading the manuscript and offering insightful suggestions. He knows dogs as well as anyone else. Finally, we would like to thank our three anonymous reviewers for exceptionally helpful feedback.

NOTES

CHAPTER 1: IMAGINING DOGS IN A WORLD WITHOUT HUMANS

1. Weisman, *World without Us*, p. 5.
2. Ibid., 44.
3. Jonathon Turnbull, a graduate student at the University of Cambridge who is studying "the 'resurgent ecologies' of Chernobyl," focusing on dogs, told us that it is not a barren landscape where the dogs are totally on their own. Phone conversation, April 23, 2020.
4. Heid, "How Dogs Would Fare without Us," 60–65.
5. Ibid., 60.
6. Ibid., 64.
7. Ibid., 64.
8. Ibid., 63.
9. Ibid., 65.
10. Ibid., 62.
11. Ibid., 62.
12. Ibid., 63.
13. Ibid., 64.
14. Ibid., 64.
15. Spotte, *Societies of Wolves and Free-ranging Dogs*, 192.

CHAPTER 2: THE STATE OF DOGS

1. Michael Fox's 1975 *The Wild Canids* is a notable exception. He devotes two chapters to free-ranging dogs. David Macdonald and Claudio Sillero-Zubiri's *The Biology and Conservation of Wild Canids* doesn't include a chapter on the dog. However, the "domestic dog" is mentioned in several places in a chapter on infectious diseases. So, at least it appears in the index.

2. Wayne and O'Brien, "Allozyme divergence within the Canidae," 339.
3. The exact number of canid species is under debate. Castelló's *Canids of the World* lists thirty-seven extant canid species. Some sources say there are only thirty-four.
4. Macdonald and Sillero-Zubiri, *Biology and Conservation of Wild Canids*, 6.
5. Bekoff and Wells, "Social Ecology and Behavior of Coyotes."
6. Freedman et al., "Genome Sequencing Highlights the Dynamic Early History of Dogs."
7. See Frantz et al., "Genomic and archaeological evidence suggest a dual origin of domestic dogs" on the likelihood of several domestication events; and Bergström et al., "Origins and genetic legacy of prehistoric dogs" for presentation of data in support of a single domestication event.
8. Jensen et al., "Genetics of How Dogs Became Our Social Allies," 334.
9. Amy Woodyatt, "Is it a dog or is it a wolf?" CNN, November 27, 2019, accessed April 14, 2020, https://www.cnn.com/travel/article/frozen-puppy-intl-scli-scn/index.html.
10. Sober, *Nature of Selection.*
11. Morey, *Dogs: Domestication and the Development of a Social Bond*, 67.
12. See Wilkins, Wrangham, and Fitch, "The 'Domestication Syndrome' in Mammals: A Unified Explanation Based on Neural Crest Cell Behavior and Genetics," for a comprehensive list of morphological traits modified in domesticated mammals.
13. Daniels and Bekoff, "Domestication, Exploitation, and Rights," 354.
14. Price, "Behavioral Aspects of Animal Domestication"; and Daniels and Bekoff, "Domestication, Exploitation, and Rights."
15. Despite being nearly everywhere we are, dogs remain surprisingly enigmatic. The truth is that nobody really knows exactly how many dogs there are in the world, nor does anyone have a comprehensive and complete picture of how and where dogs live. What we have, instead, are rough estimates, many of which capture only one segment of the world's dogs. One reason for the sketchy picture of global dog populations is that there is no single organization or governmental

agency responsible for counting dogs. Taxonomists and biologists certainly don't perform population surveys for dogs, as they do for many species of wild animals. Dogs are not on any conservation watch lists, except as threats to wild species. Instead, we have a hodgepodge of data points, often collected with some instrumental purpose in mind.

For example, data on stray dogs are collected by the World Health Organization for the express purpose of tracking and addressing the public health threat of rabies. Data on homed ("pet") dogs are piecemeal; population counts are available only in countries where pet keeping is popular or where "sheltering" is organized and aggressive. Data are typically gathered through surveys conducted by industry groups or are collected within shelters or by animal welfare organizations. Estimates from these surveys can vary by as much as 15 percent or more. Numbers of dogs might be overestimated to normalize cultural practices and expand the reach of industry. On the other hand, numbers might also be underestimated, to downplay pet dog homelessness or the killing of dogs in shelters.

If we could chart the trajectory of growth for dogs on the planet, from the early days when dogs were first speciating some 15,000 to 40,000 years ago up through the present, it would show relatively slow and steady growth from the point or points of dogs' early emergence. There would be a steep upward curve at some point over the past several hundred years, when intensive artificial selection for breeds heated up, followed by an even sharper spike in growth over the past decade or two, as the popularity of pet keeping around the world has mushroomed. Since reliable data for dogs are not available, this chart would be based largely on educated guesswork.

16. Hal Herzog estimates a ratio of 1:7.5, based on there being a billion dogs on the planet. Herzog, "Is a Love of Dogs Mostly a Matter of Where You Live?" *Psychology Today*, accessed April 14, 2020, https://www.psychologytoday.com/us/blog/animals-and-us/201908/is-love-dogs-mostly-matter-where-you-live.
17. Although numbers of dogs are available for many of the most populous countries, the data are often inconclusive, and this means that

comparisons among different countries and regions are fraught with error. For example, numbers of dogs are sometimes counted by industry groups, who would have reason to inflate their counts, or by nonprofits, who might be inclined to inaccuracies in the other direction.

18. Drawn from Hal Herzog's analysis of the Euromonitor data. Ibid.

Andrew Rowan, who has been tracking dog population demographics for years, provided the following details on dog demographics in emails sent on March 5 and July 29, 2019:

> The number of dogs in the world most likely correlates quite closely with the number of people and the formula one dog for every ten people seems to be about right for the global dog population (i.e., around 700 million dogs). Matthew Gompper projected a population of 1 billion dogs but I believe that is high.
>
> In the past forty years, the relative number of dogs (per 1000 people) has not changed much in most of the developed countries. For example, Sweden has had 70 to 80 dogs per 1000 people since 1980, the UK has had around 145 dogs per 1000 and the US has had around 225 per 1000. Globally, the number of dogs per 1000 people ranges from 2 to 3 (Saudi Arabia), to 50 to 100 (South Asia) to 250–400 the Philippines and many other Pacific Islands to 800 (rural Chile). 800 is the highest I have found so far.
>
> There are some countries that have seen changes in the relative dog population. For example, Japan has seen the number of dogs grow from 20 to 90 dogs per 1000 people over the past thirty years.
>
> Unfortunately, most of the data we have on dog populations is not particularly robust. The two main surveys in the US (the APPA and the AVMA surveys) report dog population estimates that differ by ten to fifteen percent (APPA has been consistently higher for the past fifteen years). By the way, there are around 400 million 'private' dogs (relatively controlled) and 300 million street dogs who live on the streets and are free to do what they want. Observations from Asia, Africa and Latin America indicate that most street dogs are 'owned' in that a particular household is recognized as being associated with a particular dog and providing some food for the dog. It is a mistake to view street dogs as 'strays.'"

19. This estimate is from the American Pet Products Association and is a very rough estimate. Nobody really knows how many dogs there are in the U.S. or in any other country.
20. Most estimates of the numbers of free-ranging dogs globally fall within the range of 75 percent to 85 percent, but they can vary dramatically. Morey, for example, believes that about 15 percent of the world's dogs are homed (Morey, *Dogs: Domestication and the Development of a Social Bond*), while Andrew Rowan believes the percentage to be closer to 50 percent. Rowan's estimate includes dogs who are "homeless" and living in the shelter system. Andrew says: "At any point in time, the proportion of homed dogs in shelters in North America, Europe, Australia and New Zealand is less than 1% (probably around 0.1%) of the homed dog population. An even smaller proportion of the street/community dogs are in shelters." Andrew Rowan, email communication to Marc Bekoff, December 18, 2019.
21. Morey, *Dogs: Domestication and the Development of a Social Bond*, 31.
22. Jensen, *Behavioural Biology of Dogs*, 145.
23. There is also the matter of "owned" and "owner" in talking about humans and their dogs. Animal advocates often bristle at the use of these words, and the language people use to talk about dogs certainly reveals some morally problematic assumptions. Yet many people who use the possessive to refer to their dog, as in "my dog Bella" or "our dog Rufus," do so lovingly and without any grand and evil plan about humans exploiting or commodifying dogs.
24. On the language of stray and street dogs, Arnold Arluke and Kate Atema write: "Rather than refer to these dogs as 'stray' or 'street,' we prefer the term ranging because the former implies that they are truly homeless, without any human interest in or assumption of responsibility for them. Anecdotal and survey data suggest otherwise for many of these animals." Arluke and Atema, "Roaming Dogs." https://doi.org/10.1093/oxfordhb/9780199927142.013.9.
25. Francis, *Domesticated*, 34.
26. Daniels and Bekoff, "Feralization."
27. See Gamborg et al., "De-Domestication."

28. Castelló, *Canids of the World*, 113.
29. Email communication to Marc Bekoff, December 25, 2020. For more on dingoes, see Bradley Purcell's *Dingo*; Bradley Smith's *The Dingo Debate: Origins, Behaviour and Conservation*; and Pat Shipman's "What the dingo says about dog domestication."
30. Email communication to Marc Bekoff, December 26, 2020.
31. On the concept and history of "breed," see the interesting history by Worboys, Strange, and Pemberton, *The Invention of the Modern Dog: Breed and Blood in Victorian Britain.*
32. Mark Derr, "Shifting Perspectives on How Dogs Came to Be Dogs," *Psychology Today*, September 23, 2019, accessed April 15, 2020, https://www.psychologytoday.com/us/blog/dogs-best-friend/201909/shifting-perspectives-how-dogs-came-be-dogs. Genetic analyses of dogs may pose challenges to breed conversations. People routinely mislabel breeds, with one of the most common mistakes being the mislabeling of dogs as pit bull type. Dogs who are labeled and sold as purebred often are not; dog owners who do DNA analysis of their so-called purebred dog often get an unpleasant surprise when they get the test results.
33. See Michael Brandow, *A Matter of Breeding.*
34. See "Pets by the Numbers," *Animal Sheltering*, accessed April 15, 2020, https://humanepro.org/page/pets-by-the-numbers. Although sheer numbers of purebred dogs may be increasing, the percentage of purebreds relative to mixed breeds may be decreasing slightly in some places, including the US.
35. Marc Bekoff, "Dog Breeds Don't Have Distinct Personalities," *Psychology Today*, https://www.psychologytoday.com/us/blog/animal-emotions/201901/dog-breeds-dont-have-distinct-personalities.

CHAPTER 3: THE SHAPE OF THE FUTURE

1. Stearns, "Trade-offs in life history evolution," 4.
2. There are some notable exceptions, particularly in research on free-ranging dog populations. For example, Boitani, Francisci, Ciucci, and Andreoli include a section on reproduction and life histories in a paper on the ecology and behavior of feral dogs in Italy ("Population biology and ecology of feral dogs in central Italy").

3. For example, see Asher et al. "Inherited defects in pedigree dogs. Part 1: disorders related to breed standards" and Edmunds et al. "Dog breeds and body conformations with predisposition to osteosarcoma in the UK: a case-control study."
4. See Bekoff, Daniels, and Gittleman, "Life history patterns and the comparative social ecology of carnivores."
5. Jonathan Losos's *Improbable Destinies* is a fascinating look at convergent evolution. He writes about "evolutionary regularities," or rules that seem to hold, for the most part, such as Bergmann's and Allen's rules. But evolutionary lineages often take different paths in response to the same pressure. So, hot climates might lead to smaller as well as larger body sizes.
6. Cooke, Eigenbrod, and Bates, "Projected losses of global mammal and bird ecological strategies."
7. Ibid.
8. Hemmer, *Domestication*, 26.
9. See Lark et al., "Genetic architecture of the dog."
10. Bryce and Williams, "Comparative locomotor costs of domestic dogs."
11. Ibid.
12. See, for example, Andersson, "Were there pack-hunting canids in the Tertiary, and how can we know?"
13. See Fawcett et al., "Consequences and Management of Canine Brachycephaly in Veterinary Practice: Perspectives from Australian Veterinarians and Veterinary Specialists." They concede, "Whether the recent selection of dogs with progressively shorter and wider skulls has reached physiological limits is controversial." But they conclude that some dog breeds are at a breaking point.
14. Kaminski et al. "Evolution of facial muscle anatomy in dogs."
15. Ample data suggest a role for oxytocin in human-dog relationships, but it is unclear still exactly what this relationship is. On "oxytocin feedback loops," see contradictory results by Sarah Marshall-Pescini and her colleagues. Marshall-Pescini et al., "The Role of Oxytocin in the Dog–Owner Relationship."
16. Perry et al. "Epidemiological study of dogs with otitis externa."
17. Spotte, *Societies of Wolves and Free-ranging Dogs*, 55.

18. Parker et al. "The bald and the beautiful."
19. Hemmer, *Domestication.*
20. If you are interested in going down the rabbit hole on weather, climate, and thermoregulation, see da Silva and Campos Maia, *Principles of Animal Biometeorology.* On page 119, they talk about potential advantages of dark pigmentation at lower latitudes.

CHAPTER 4: FOOD AND SEX

1. Facultative carnivores—including domestic dogs—can survive well on a diet in which many of their calories are derived from a potpourri of sources. An omnivore can get nutrients and food energy from both animals and plants and sometimes even items that we wouldn't think of as food. Coyotes, for example, have incredibly diverse meal plans and one study found that the coyotes whose poop they analyzed had a diet that included rubber balls, rubber boots, gloves, cotton, and other nonfood items among the sixty or so items that were found in their scats. Field biologists typically categorize the "omnivorous carnivore" as a species in which meat constitutes less than about half of the diet. "Carnivorous carnivores" rely on meat for more than half their diet.
2. See Butler, Brown, and Du Toit, "Anthropogenic Food Subsidy to a Commensal Carnivore: The Value and Supply of Human Faeces in the Diet of Free-Ranging Dogs."
3. Spotte, *Societies of Wolves and Free-ranging Dogs*, 143.
4. Daniels, "Conspecific scavenging by a young domestic dog,"
5. For example, see Young et al. "Is Wildlife Going to the Dogs? Impacts of Feral and Free-roaming Dogs on Wildlife Populations."
6. Stephen Spotte, for example, includes this observation: "Three dogs in St. Louis were seen chasing squirrels in a park on 61 occasions, all pursuits unsuccessful." Spotte, *Societies of Wolves and Free-Ranging Dogs*, 143. Marc observed some of these chases and knew they were doomed from the start because many seemed to be instances of play, not hunting.
7. See Gompper *Free-Ranging Dogs &Wildlife Conservation.*

8. Our copyeditor remarked, "This example gave me a flashback to a book I read as a kid: *Desert Dog*, by Jim Kjelgaard. It's about a greyhound who is abandoned in the desert and must adapt or perish." Kjelgaard's fictional book, which follows the adventures of Tawny, Sable, and Brutus, is a remarkably convincing account of dogs forming packs, making friends, squabbling, and learning to hunt and survive without human care.
9. J. David Henry, "Red Fox."
10. Ritchie et al., "Dogs as predators and trophic regulators," 58.
11. Samuel et al., "Fears from the past?"
12. Sarkar, Sau, and Bhadra, "Scavengers can be choosers," 38.
13. Macdonald, Creel, and Mills, "Canid Society," 85.
14. Ibid., 87.
15. Researchers studying wolves agree that inbreeding is relatively rare. Inbreeding depression is nevertheless recognized as a serious problem in some wolf populations, particularly those that are geographically isolated, such as the wolves on Isle Royale. Wolves have been under such intense pressure from human predation that the number of breeding individuals is often quite small, and genetic bottlenecks can occur. When significant inbreeding occurs, packs are less likely to survive. See, for further discussion, Bensch et al., "Selection for Heterozygosity Gives Hope to a Wild Population of Inbred Wolves"; Ralls, Harvey, and Lyles, "Inbreeding in natural populations of birds and mammals"; Lockyear et al., "Retrospective Investigation of Captive Red Wolf Reproductive Success in Relation to Age and Inbreeding."

    In emails to Marc, renowned wolf researchers L. David Mech, Douglas Smith, and Rick McIntyre all agreed that inbreeding among wild wolves is relatively rare (personal emails on April 12, 2020).
16. Kopaliani et al., "Gene flow between wolf and shepherd dog populations in Georgia (Caucasus)." Tibetan mastiffs, who were bred as flock guardians on the Tibetan Plateau in the Himalayas, are an interesting example of heterosis—interbreeding with benefits. The Tibetan mastiffs interbred with Tibetan gray wolves, which is why they are

particularly well adapted to their high-altitude ecological niche. Signore et al., "Adaptive Changes in Hemoglobin Function."

17. Daniels and Bekoff, "Population and Social Biology of Free-Ranging Dogs, *Canis familiaris*," 758.
18. Majumder et al., "Denning habits of free-ranging dogs," 2.
19. Bonanni and Cafazzo, "The Social Organization of a Population of Free-Ranging Dogs," 84.
20. Pal, "Parental care in free-ranging dogs, *Canis familiaris*," 31.
21. Paul and Bhadra, "The Great Indian Joint Families of Free-Ranging Dogs," 1.
22. Ibid., 13.
23. Pal, Roy, and Ghosh, "Pup rearing."
24. Pongrácz and Sztruhala, "Forgotten, But Not Lost," 1.
25. Darcy Morey raises an intriguing question. He describes dogs as "r-strategists" when associated with people, in contrast to wolves who are K-strategists. Biologists Edward Wilson and Robert MacArthur introduced the concepts of r- and K- selection in their 1967 book *The Theory of Island Biogeography*, to describe two different evolutionary life-history strategies that have evolved in response to different ecological pressures. In unstable environments, the best strategy is to produce many offspring and not invest too much in them. Characteristic features of r-strategists are early maturity, large numbers of offspring, little to no parental care, and large reproductive effort. K-strategies tend to evolve in stable environments. K-strategists display delayed reproduction, small numbers of young, and high levels of parental care. Is Morey right, and what would the implications be for posthuman dogs? Would dogs shift toward a K-strategy, if humans weren't in the picture? How long would it take dogs to shift from r- to K-strategy?
26. Some data suggest that larger breeds of dogs reach sexual maturity later than smaller breeds.
27. Mech and Boitani, *Wolves: Behavior, Ecology, and Conservation.*
28. Packard et al. "Causes of Reproductive Failure"; and Sands and Creel, "Social dominance, aggression and faecal glucocorticoid levels."
29. Bonanni and Cafazzo, "Social Organization of a Population of Free-Ranging Dogs," 82.

30. Brisbin and Risch, "Primitive dogs, their ecology and behavior."
31. McIntyre et al., "Behavioral and ecological implications of seasonal variation."
32. Daniels and Bekoff, "Population and Social Biology of Free-Ranging Dogs," 757.
33. Sidorovich et al., "Litter size, sex ratio, and age structure of gray wolves."
34. Mech, *The Wolf.*
35. See, for example, Inoue et al., "A current life table and causes of death for insured dogs in Japan."
36. Spotte, *Societies of Wolves and Free-ranging Dogs*, 191.
37. Paul et al., "High early life mortality in free-ranging dogs."
38. Ibid., 1.
39. Daniels and Bekoff reported that mortality of free-ranging dogs was relatively high in early life, although it is hard to determine this with certainty because dog carcasses were rarely found. For the litters of five females, 7 percent of the pups died; overall survival rate to four months of age was 34 percent. Daniels and Bekoff, "Population and Social Biology of Free-Ranging Dogs," 757. From Pal, "Parental care in free-ranging dogs, *Canis familiaris*": "Puppy mortality rates of 63% by the age of 3 months was in approximate agreement with the results of previous studies in free-roaming domestic dogs by Scott and Causey (1973), Daniels and Bekoff (1989) and Pal (2001)."

CHAPTER 5: FAMILY, FRIEND, AND FOE

1. Leyhausen, "The Communal Organization of Solitary Animals."
2. Scott and Fuller, *Genetics and the Social Behavior of the Dog.* Depending on the situation in which a pup is reared, it is possible that the socialization period could extend to upwards of twelve to fourteen weeks of age. Freedman, King, and Elliot, "Critical Period in the Social Development of Dogs."
3. Socialization is not the same as habituation. Habituation is generally characterized as a decreased response to repeated stimuli.
4. See Pierce, *Run, Spot, Run*; and Bekoff and Pierce, *Unleashing Your Dog.*

5. Bekoff and Wells, "Social Ecology and Behavior of Coyotes"; Burrows, *Wild Fox*; and Van Lawick-Goodall, *Innocent Killers.*
6. Fox, *Behaviour of Wolves, Dogs, and Related Canids.*
7. Bekoff, "Socialization in mammals with an emphasis on non-primates."
8. Fox, *Behavior of Wolves, Dogs, and Related Canids.*
9. Faragó et al., "Dogs' Expectation about Signalers' Body Size."
10. Riach, Asquith, and Fallon, "Length of time domestic dogs (*Canis familiaris*) spend smelling urine."
11. "A castrated dog will still urine-mark, using the characteristic male leg-lift posture, but it will do so less often." Ian Dunbar, Neutering Fact Sheet, *Modern Dog Magazine*, accessed April 15, 2020, https://moderndogmagazine.com/articles/neutering-fact-sheet/255.
12. Macdonald and Carr, "Variation in dog society," 333.
13. Ibid., 326.
14. Also see Boitani et al., "The ecology and behavior of feral dogs: A case study from central Italy."
15. Beck, *Ecology of Stray Dogs.*
16. Macdonald and Carr, "Variation in dog society," 335.
17. Ibid., 336.
18. Daniels and Bekoff, "Population and Social Biology of Free-Ranging Dogs," 754.
19. Ibid., 753. Miternique and Gaunet, in "Coexistence of Diversified Dog Socialities and Territorialities in the City of Concepción, Chile," present a very interesting discussion. "New forms of sociality were also evidenced, with dogs exhibiting intermediate degrees of sociality between the *pet* and *stray dog* categories. We postulate that this unique diversity of sociospatial positioning and level of adjustment (e.g., dogs using crosswalks either alone or with people) is made possible by the city's specific human culture and range of urban areas. The dog species thus exhibits a considerable potential for social and spatial adjustment. The fact that it depends on the spatial layout and human culture of their environment explains the presence of dogs wherever humans are."
20. Bekoff, "Mammalian Dispersal."
21. Comments submitted to our Princeton University Press editor by an anonymous reviewer of this manuscript.

22. Feddersen-Petersen, "Social Behaviour of Dogs," 100–11.
23. Bonanni and Cafazzo, "Social Organization of a Population of Free-Ranging Dogs," 78.
24. Ibid., 70, 78.
25. Schenkel, "Expressions Studies on Wolves."
26. Burt, "Territoriality and home range concepts."
27. Beck, "Ecology of 'Feral' and Free-Roving Dogs."
28. Spotte, *Societies of Wolves and Free-ranging Dogs*, 112.
29. Macdonald and Carr, "Variation in dog society"; and Boitani et al., "Population biology and ecology of feral dogs."
30. Gompper, *Free-Ranging Dogs and Wildlife Conservation*, 27. Here he is drawing on work by Luigi Boitani, David Macdonald, and others.
31. Bonnani and Cafazzo, "Social Organization of a Population of Free-Ranging Dogs," 90.
32. Daniels and Bekoff, "Spatial and Temporal Resource Use," 306.
33. Ibid., 308.
34. See Vanak and Gompper, "Dietary niche separation between sympatric free-ranging domestic dogs and Indian foxes in central India"; Vanak et al., "Top-dogs and underdogs: Competition between dogs and sympatric carnivores."
35. On dogs as members of carnivore guilds, see Vanak et al. "Top-dogs and underdogs: Competition between dogs and sympatric carnivores."
36. Ibid., 69.
37. Current dogs are already in sharp competition with wild species and have significant impacts on ecosystems. A group of Australian researchers has been cataloging the impacts of domestic dogs on wildlife around the globe, particularly on vertebrate species with a conservation status of "threatened." They summarized some of their findings in 2017 in the journal *Biological Conservation.* They found that domestic dogs—including feral, free-ranging, and owned pet dogs—have contributed to 11 vertebrate extinctions and are a known or potential threat to at least 188 threatened species worldwide. "Predation by dogs was the most frequently reported impact, followed by disturbance, disease transmission, competition, and hybridisation." Doherty et al., "The global impacts of domestic dogs on threatened vertebrates," 56.

A 2019 *Washington Post* article about the problem of dogs killing wildlife in Brazil focused on the work done by this team of researchers:

> It's a question more researchers are beginning to ask in a country where there are more dogs than children—and where dogs are quickly becoming the most destructive predator. They're invading nature preserves and national parks. They're forming packs, some 15 dogs strong, and are hunting wild prey. They've muscled out native predators such as foxes and big cats in nature preserves, outnumbering pumas 25 to 1 and ocelots 85 to 1. . . .
>
> The researchers, based in Australia, convicted dogs in the extinction of 11 species and declared them the third-most-damaging mammal, behind only cats and rodents.
>
> The International Union for Conservation of Nature maintains a list of animals whose numbers dogs are culling. There are 191, and more than half are classified as either endangered or vulnerable. They range from lowly iguanas to the famed Tasmanian devil, from doves to monkeys, a diversity of animals with nothing in common beyond the fact that dogs enjoy killing them. In New Zealand, the organization reported, a single German shepherd once did in as many as 500 kiwis—and that was the conservative estimate.
>
> Terrence McCoy, "The dog is one of the world's most destructive mammals. Brazil proves it," *Washington Post*, August 20, 2019, accessed April 15, 2020. https://www.washingtonpost.com/world/the_americas/the-dog-is-one-of-the-worlds-most-destructive-mammals-brazil-proves-it/2019/08/19/c37a1250-a8da-11e9-8733-48c87235f396_story.html.

38. Boydston et al. "Canid vs. canid:."

CHAPTER 6: THE INNER LIVES OF POSTHUMAN DOGS

1. The word "intelligence" refers to the ability of an individual to acquire knowledge and to use it to adapt to different situations and do what's needed to accomplish various tasks. Howard Gardner introduced the idea of "multiple intelligences" in his 1983 book *Frames of Mind*. Gardner's hypothesis was that intelligence is not a generalized capacity but rather involves several different "modalities"—different ways of processing information. Some people, for example, are visu-

ally oriented while others are verbally oriented. Dogs, too, have a range of different "smarts" and a unique profile of intellectual skills that work for them.

2. Bekoff and Pierce, *Unleashing Your Dog.*
3. Cauchoix, Chaine, and Barragan-Jason, "Cognition in Context"; Szabo, Damas-Moreira, and Whiting, "Can Cognitive Ability Give Invasive Species the Means to Succeed?"
4. Bhattacharjee Sau, and Bhadra, "Free-Ranging Dogs Understand Human Intentions and Adjust Their Behavioral Responses Accordingly," focuses on free-ranging dogs following human pointing gestures. The study also provides an overview of the research on homed dogs and wolves.
5. See, for instance, Belger and Bräuer, "Metacognition in dogs: Do dogs know they could be wrong?"
6. Lazzaroni et al., "The role of life experience in affecting persistence."
7. Benson-Amram and Holekamp, "Innovative problem solving," 4087.
8. Marshall-Pescini et al., "Does training make you smarter?"
9. Marshall-Pescini et al., "Importance of a species' socioecology."
10. Range et al., "Wolves lead and dogs follow."
11. Drea and Carter, "Cooperative problem solving in a social carnivore."
12. Marshall-Pescini et al., "Importance of a species' socioecology," 11793.
13. Researchers in ethology and in comparative psychology approach the study of animal personality very differently, using different assumptions and different models. But despite differences, both embrace the idea that animals have personalities. See Weiss, "Personality Traits: A View from the Animal Kingdom" and Jones and Gosling, "Temperament and personality in dogs (*Canis familiaris*): A review and evaluation of past research."
14. Bremner-Harrison, Prodohl, and Elwood, "Behavioural trait assessment as a release criterion."
15. Santicchia et al., "The price of being bold?"
16. Seyle, *The Stress of Life.*
17. Vindas et al., "How do individuals cope with stress?" 1524. See also Koolhaas et al., "Coping styles in animals: current status in behavior and stress-physiology."

18. For example, Hiby, Rooney, and Bradshaw, “Behavioural and physiological responses of dogs.”
19. Horváth et al., “Three different coping styles in police dogs.”
20. Špinka, Newberry, and Bekoff, “Mammalian play.”
21. Brand, *Hidden World of the Fox.*
22. Maglieri et al., “Levelling playing field.”
23. Altmann, *Baboon Mothers and Infants.* Altmann characterized baboon mothers as laissez-faire or restrictive.

CHAPTER 7: DOOMSDAY PREPPING

1. *Doomsday Preppers*, National Geographic, https://www.nationalgeographic.com.au/tv/doomsday-preppers/.
2. Emily S. Rueb and Niraj Chokshi, “Labradoodle Creator Says the Breed is His Life’s Regret,” *New York Times*, September 25, 2019, accessed April 15, 2020, https://www.nytimes.com/2019/09/25/us/labradoodle-creator-regret.html.
3. Dogs also breed with coyotes, wolves, and jackals in the wild.
4. Another complicating factor is that desexing seems to have complex and very mixed health implications for dogs. The AVMA notes that there are competing risks and benefits to spaying/neutering: “For canine patients, due to the varied incidence and severity of disease processes, there is no single recommendation that would be appropriate for all dogs. Developing recommendations for an informed case-by-case assessment requires an evaluation of the risks and benefits of spay/neuter, including its potential effects on neoplasia, orthopedic disease, reproductive disease, behavior, longevity, and population management. However, many factors other than neuter status play an important role in these outcomes, including breed, sex, genetics, lifestyle, and body condition.” American Veterinary Medical Association, “Elective spaying and neutering of pets,” accessed April 15, 2020, https://www.avma.org/resources-tools/animal-health-and-welfare/elective-spaying-and-neutering-pets.
5. A few of the many effects on dogs so far: large numbers of shelter dogs were adopted or fostered, as people were ordered to stay at home for

an extended period of time; large numbers of homed dogs have been abandoned or relinquished to shelters, both because of unfounded fears that dogs may carry the virus and because the economic fallout of the pandemic has meant that many people can't afford to feed and care for a dog they already own; for street and feral dogs, lockdown orders have meant fewer people providing handouts and providing food subsidies.

6. We aren't using the term "euthanasia" because this concept suffers from profound ambiguity and is often used, in relation to dogs, with almost no ethical nuance. "Euthanasia" should only rightly be used to describe the hastening of death, usually with an injection of sodium pentobarbital, of a specific individual dog and in response to unrelenting and untreatable suffering because of illness or injury. The "putting down" of healthy dogs, whether in shelters at the request of society or in veterinary clinics at the request of individual pet owners, is *not* euthanasia. Similarly, killing massive numbers of healthy dogs because we thought they might potentially suffer when we are gone is also *not* euthanasia.
7. Kean, *Great Dog and Cat Massacre*. Campbell's *Bonzo's War* also recounts this horrific event.

CHAPTER 8: WOULD DOGS BE BETTER OFF WITHOUT US?

1. Pierce, *Run, Spot, Run*.
2. Brooks, *The Grass Library: Essays*, 15.
3. Pierce, "Beyond Humans: Dog Utopia or Dog Dystopia?" *Psychology Today* (blog), October 18, 2018, https://www.psychologytoday.com/ca/blog/all-dogs-go-heaven/201810/beyond-humans-dog-utopia-or-dog-dystopia.
4. See Bekoff and Pierce, *Animals' Agenda*; and Bekoff and Pierce, *Unleashing Your Dog*.

BIBLIOGRAPHY

Abrantes, Roger. *The Evolution of Canine Social Behavior.* Naperville, IL: Wakan Tanka Publishers, 1997.

Adda, Marco. "Free-Ranging Dogs for a Multispecies Landscape: A Paradigm Shift in an Essential Piece of Human-Animal Coexistence." In *Anthrozoology Studies. Thinking beyond Boundaries*, edited by I. Frasin, G. Bodi, C. Dinu Vasiliu, 117–34. Bucharest: Pro Universitaria, 2020. (In Romanian) ISBN: 978-606-26-1212-2.

Allan, James R., James E. M. Watson, Moreno Di Marco, Christopher J. O'Bryan, Hugh P. Possingham, Scott C. Atkinson, and Oscar Venter, "Hotspots of human impact on threatened terrestrial vertebrates." *PLOS Biology* 17, no. 3 (2019): e3000158. https://doi.org/10.1371/journal.pbio.3000158.

Altmann, Jeanne. *Baboon Mothers and Infants.* Cambridge: Harvard University Press, 1980.

American Veterinary Medical Association. "Elective spaying and neutering of pets." Accessed April 15, 2020. https://www.avma.org/resources-tools/animal-health-and-welfare/elective-spaying-and-neutering-pets.

Andersson, K. "Were there pack-hunting canids in the Tertiary, and how can we know?" *Paleobiology* 31, no. 1 (2005): 56–72.

Arluke, Arnold, and Kate Atema. "Roaming Dogs." In *The Oxford Handbook of Animal Studies*, edited by Linda Kalof. Oxford Handbooks Online, July 2015. https://doi.org/10.1093/oxfordhb/9780199927142.013.9.

Asher, Lucy, Gillian Diesel, Jennifer F. Summers, Paul D. McGreevy, Lisa M. Collins. "Inherited defects in pedigree dogs. Part 1: disorders related to breed standards." *Veterinary Journal* 182 (2009): 402–11. https://doi.org/10.1016/j.tvjl.2009.08.033. PMID: 19836981.

Bar-On, Yinon M., Rob Phillips, and Ron Milo. "The biomass distribution on Earth." *Proceedings of the National Academy of Sciences* 115, no. 25 (2018): 6506–11. https://doi.org/10.1073/pnas.1711842115.

Barrett, Lisa P., Lauren Stanton, Sarah Benson-Amram. "The cognition of 'nuisance' species." *Animal Behaviour* 147 (2019): 167–77. https://doi.org/10.1016/j.anbehav.2018.05.005.

Bartos, Ludek, Jitka Bartosová, Helena Chaloupková, Adam Dusek, Lenka Hradecká, and Ivona Svobodová. "A sociobiological origin of pregnancy failure in domestic dogs." *Scientific Reports* 6 (2016): 22188. https://doi.org/10.1038/srep22188.

Baum, S., S. Armstrong, T. Ekenstedt, O. Häggström, R. Hanson, K. Kuhlemann, M. Maas, J. Miller, M. Salmela, A. Sandberg, K. Sotala, P. Torres, A. Turchin, and R. Yampolskiy. "Long-term trajectories of human civilization." *Foresight* 21, no. 1 (2019): 53–83. https://doi.org/10.1108/FS-04-2018-0037.

Beck, Alan M. "The Ecology of 'Feral' and Free-Roving Dogs in Baltimore." In *The Wild Canids: Their Systematics, Behavioral Ecology and Evolution*, edited by Michael W. Fox, 380–90. New York: Litton, 1975.

———. *The Ecology of Stray Dogs: A Study of Free-Ranging Urban Animals*. West Lafayette, IN: Purdue University Press, 1973.

Bekoff, Marc. *Canine Confidential: Why Dogs Do What They Do.* Chicago: Chicago University Press, 2018.

———. "Dog Breeds Don't Have Distinct Personalities," *Psychology Today*, https://www.psychologytoday.com/us/blog/animal-emotions/201901/dog-breeds-dont-have-distinct-personalities.

———. "Dumping the dog domestication dump theory once and for all," *Psychology Today* (blog), November 11, 2018. https://www.psychologytoday.com/us/blog/animal-emotions/201811/dumping-the-dog-domestication-dump-theory-once-and-all.

———. "Mammalian Dispersal and the Ontogeny of Individual Behavioral Phenotypes." *American Naturalist* 111 (1977): 715–32.

———. "Socialization in mammals with an emphasis on non-primates." In *Primate bio-social development*, edited by S. Chevalier-Skolnikoff and F. E. Poirier, 603–36. New York: Garland Publishers, 1977.

Bekoff, Marc, and John A. Byers. "The Development of Behavior from Evolutionary and Ecological Perspectives in Mammals and Birds." In *Evolutionary Biology*, edited by M. K. Hecht, B. Wallace, and G. T. Prance, 215–86. Springer, Boston, MA: 1985.

Bekoff, Marc, Thomas J. Daniels, and John L. Gittleman. "Life history patterns and the comparative social ecology of carnivores." *Annual Review of Ecology and Systematics* 15 (1984): 191–232.

Bekoff, Marc, Judy Diamond, and Jeffry B. Mitton. "Life-history patterns and sociality in canids: Body size, reproduction, and behavior." *Oecologia* 50 (1981): 386–90.

Bekoff, Marc, and Jessica Pierce. *The Animals' Agenda: Freedom, Compassion, and Coexistence in the Human Age*. Boston: Beacon Press, 2017.

———. *Unleashing Your Dog: A Field Guide to Giving Your Canine Companion the Best Life Possible*. Novato, CA: New World Library, 2019.

Bekoff, Marc, and Michael C. Wells. "Social Ecology and Behavior of Coyotes." *Advances in the Study of Behavior* 16 (1986): 251–338. https://animalstudiesrepository.org/cgi/viewcontent.cgi?article=1036&context=acwp_ena.

Belger, Julia, and Juliane Bräuer. "Metacognition in dogs: Do dogs know they could be wrong?" *Learning and Behavior* 46 (2018): 398–413. https://doi.org/10.3758/s13420-018-0367-5.

Belo, V. S., G. L. Werneck, E. S. da Silva, D. S. Barbosa, C. J. Struchiner. "Population Estimation Methods for Free-Ranging Dogs: A Systematic Review." *PLOS ONE* 10, no. 12 (2015): e0144830. https://doi.org/10.1371/journal.pone.0144830.

Bensch, Staffan, Henrik Andrén, Bengt Hansson, Hans Chr. Pedersen, Håkan Sand, Douglas Sejberg, Petter Wabakken, Mikael Åkesson, and Olof Liberg. "Selection for Heterozygosity Gives Hope to a Wild Population of Inbred Wolves." *PLOS ONE* 1, no. 1 (2006): e72. https://doi.org/10.1371/journal.pone.0000072.

Benson-Amram, Sarah, Geoff Gilfillan, and Karen McComb. "Numerical assessment in the wild: insights from social carnivores." *Philosophical Transactions of the Royal Society B* 373 (2017): 20160508. http://dx.doi.org/10.1098/rstb.2016.0508.

Benson-Amram, Sarah, and Kay E. Holekamp. "Innovative problem solving by wild spotted hyenas." *Proceedings of the Royal Society B: Biological Sciences* (2012): 4087–95. https://doi.org/10.1098/rspb.2012.1450.

Bergström, Anders, Laurent Frantz, Ryan Schmidt, Erik Ersmark, Ophelie Lebrasseur, Linus Girdland-Flink, Audrey T. Lin, Jan Storå, Karl-Göran Sjögren, David Anthony, et al. "Origins and genetic legacy of prehistoric dogs." *Science* 370, no. 6516 (2020): 557–64. https://doi.org/10.1126/science.aba9572.

Bhattacharjee, Debottam, Sarab Mandal, Piuli Shit, Mebin George Varghese, Aayushi Vishnoi, and Anindita Bhadra. "Free-Ranging Dogs Are Capable of Utilizing Complex Human Pointing Cues." *Frontiers in Psychology* 10 (2020). https://doi.org/10.3389/fpsyg.2019.02818.

Bhattacharjee, Debottam, Shubhra Sau, and Anindita Bhadra. "Free-Ranging Dogs Understand Human Intentions and Adjust Their Behavioral Responses Accordingly." *Frontiers in Ecology and Evolution* 6 (2018). https://www.frontiersin.org/articles/10.3389/fevo.2018.00232/full.

Bielby J., G. M. Mace, O. R. P. Bininda-Emonds, M. Cardillo, J. L. Gittleman, K. E. Jones, C. D. L. Orme, and A. Purvis. "The fast-slow continuum in mammalian life history: an empirical reevaluation." *American Naturalist* 169 (2007): 748–57. https://doi.org/10.1086/516847.

Biro, Peter. A., and Judy A. Stamps. "Are animal personality traits linked to life-history productivity?" *Trends in Ecology and Evolution* 23 (2008): 361–68.

Boitani, L., and P. Ciucci. "Comparative social ecology of feral dogs and wolves." *Ethology, Ecology and Evolution* 7 (1995): 49–72.

Boitani, L., F. Francisci, P. Ciucci, and G. Andreoli. "Population biology and ecology of feral dogs in central Italy." In *The Domestic Dog*,

2nd ed., edited by James Serpell, 342–68. Cambridge: Cambridge University Press, 2017.

Boitani, Luigi, Paolo Ciucci, and Alessia Ortolani. "Behaviour and Social Ecology of Free-Ranging Dogs." In *The Behavioural Biology of Dogs*, edited by Per Jensen, 147–65. Oxfordshire, UK: CAB International, 2007.

Bonanni Roberto, Simona Cafazzo, Arianna Abis, Emanuela Barillari, Paola Valsecchi, and Eugenia Natoli. "Age-graded dominance hierarchies and social tolerance in packs of free-ranging dogs." *Behavioral Ecology* 28 (2017): 1004–20. https://doi.org/10.1093/beheco/arx059.

Bonanni, Roberto, and Simona Cafazzo. "The Social Organization of a Population of Free-Ranging Dogs in a Suburban Area of Rome: A Reassessment of the Effects of Domestication on Dogs' Behaviour." In *The Social Dog: Behaviour and Cognition*, edited by Juliane Kaminski and Sarah Marshall-Pescini, 65–104. San Diego: Elsevier, 2014.

Bonanni, Roberto, Simona Cafazzo, Paola Valsecchi, and Eugenia Natoli. "Effect of affiliative and agonistic relationships on leadership behaviour in free-ranging dogs." *Animal Behaviour* 79 (2010): 981–91.

Bonanni, Roberto, Eugenia Natoli, Simona Cafazzo, Paola Valsecchi. "Free-ranging dogs assess the quantity of opponents in inter-group conflicts." *Animal Cognition* 14 (2011): 103–15.

Bonanni, Roberto, Paola Valsecchi, and Eugenia Natoli. "Pattern of individual participation and cheating in conflicts between groups of free-ranging dogs." *Animal Behaviour* 79 (2010): 957–68.

Bostrom, Nick, and Milan M. Ćirković. *Global Catastrophic Risks*. Oxford: Oxford University Press, 2008.

Boydston, Erin E., Eric S. Abelson, Ari Kazanjian, and Daniel T. Blumstein. "Canid vs. canid: insights into coyote-dog encounters from social media." *Human-Wildlife Interactions* 12, no. 2 (2018): 233–42.

Boyko, Adam R. "The domestic dog: man's best friend in the genomic era." *Genome Biology* 12, no. 2 (2011): 216.

Bradley P. Smith, Kylie M. Cairns, Justin W. Adams, Thomas M. Newsome, Melanie Fillios, Eloïse C. Déaux, William C. H. Parr, Mike Letnic, Lily M. Van Eeden, Robert G. Appleby, et al. "Taxonomic status of the Australian dingo: the case for Canis dingo Meyer, 1793." *Zootaxa* 4564, no. 1 (2019): 173–97. https://doi.org/10.11646/zootaxa.4564.1.6.

Brand, Adele. *The Hidden World of the Fox.* New York: William Morrow, 2019.

Brandow, Michael. *A Matter of Breeding: A Biting History of Pedigree Dogs and How the Quest for Status Has Harmed Man's Best Friend.* Boston: Beacon Press, 2015.

Breck, Stewart W., Sharon A. Poessel, Peter Mahoney, and Julie K. Young. "The intrepid urban coyote: a comparison of bold and exploratory behavior in coyotes from urban and rural environments." *Scientific Reports* 9 (2019): 2104. https://doi.org/10.1038/s41598-019-38543-5.

Bremner-Harrison, S., P. A. Prodohl, and R. W. Elwood. "Behavioural trait assessment as a release criterion: boldness predicts early death in a reintroduction programme of captive-bred swift fox (*Vulpes velox*)." *Animal Conservation* 7 (2004): 313–20.

Bricker, Darrell, and John Ibbitson. *Empty Planet: The Shock of Global Population Decline.* New York: Crown, 2019.

Brisbin, I. L., and Thomas S. Risch. "Primitive dogs, their ecology and behavior: Unique opportunities to study the early development of the human-canine bond." *Journal of the American Veterinary Medical Association* 210, no. 8 (1997): 1122–26.

Brooks, David G. *The Grass Library: Essays.* Ashland, OR: Ashland Creek Press.

Bryce, Caleb M., and Terrie M. Williams. "Comparative locomotor costs of domestic dogs reveal energetic economy of wolf-like breeds." *Journal of Experimental Biology* 220 (2017): 312–21. https://doi.org/10.1242/jeb.144188.

Bubna-Littitz, Hermann. "Sensory Physiology and Dog Behaviour." In *The Behavioural Biology of Dogs*, edited by Per Jensen, 91–104. Oxfordshire, UK: CAB International, 2007.

Budaev, Sergey, Christian Jørgensen, Marc Mangel, Sigrunn Eliassen, and Jarl Giske. "Decision-Making from the Animal Perspective: Bridging Ecology and Subjective Cognition." *Frontiers in Ecology and Evolution* 7 (2019): 164. https://doi.org/10.3389/fevo.2019.00164.

Burrows, Roger. *Wild Fox.* Taplinger, 1968.

Burt, William Henry. "Territoriality and home range concepts as applied to mammals." *Journal of Mammalogy* 24 (1943): 346–52.

Butler, James R. A., Wendy Y. Brown, and Johan T. Du Toit. "Anthropogenic Food Subsidy to a Commensal Carnivore: The Value and Supply of Human Faeces in the Diet of Free-Ranging Dogs." *Animals* 8, no. 5 (2018): 67. https://doi.org/10.3390/ani8050067.

Byrne, Richard. *The Thinking Ape: The Evolutionary Origins of Intelligence.* Oxford: Oxford University Press, 1995.

Cafazzo, Simona, Roberto Bonanni, Paola Valsecchi, and Eugenia Natoli. "Social variables affecting mate preferences, copulation and reproductive outcome in a pack of free-ranging dogs." *PLOS ONE* 9 (2014): e98594. https://doi.org/10.1371/journal.pone.0098594.

Cafazzo, Simona, Paola Valsecchi, Roberto Bonanni, and Eugenia Natoli. "Dominance in relation to age, sex, and competitive contexts in a group of free-ranging domestic dogs." *Behavioral Ecology* 21, no. 3 (2010): 443–55. https://doi.org/10.1093/beheco/arq001.

Campbell, Clare. *Bonzo's War: Animals under Fire, 1939–1945.* London: Constable, 2014.

Careau, Vincent, Denis Reale, Murray M. Humphries, and Donald W. Thomas. "The pace of life under artificial selection: personality, energy expenditure, and longevity are correlated in domestic dogs." *American Naturalist* 175, no. 6 (2010): 753–58.

Carrasco, Johanna J., Dana Georgevsky, Michael Valenzuela, and Paul D. McGreevy. "A pilot study of sexual dimorphism in the head morphology of domestic dogs." *Journal of Veterinary Behavior* 9, no. 1 (2014): 43–46.

Carter, Alecia. J., William E. Feeney, Harry H. Marshall, Guy Cowlishaw, Robert Heinsohn. "Animal personality: what are behavioural ecologists measuring?" *Biological Review* 88 (2013): 465–75.

Castelló, José R. *Canids of the World.* Princeton: Princeton University Press, 2018.

Cauchoix Maxime, Alexis S. Chaine, and Galdys Barragan-Jason. "Cognition in Context: Plasticity in Cognitive Performance in Response to Ongoing Environmental Variables." *Frontiers in Ecology and Evolution* 8 (2020): 106. https://doi.org/10.3389/fevo.2020.00106.

Chu, Erin T., Missy J. Simpson, Kelly Diehl, Rodney L. Page, Aaron J. Sams, and Adam R. Boyko. "Inbreeding depression causes reduced fecundity in Golden Retrievers." *Mammalian Genome* 30 (2019): 166.

Clauset, Aaron, and Douglas H. Erwin. "The Evolution and Distribution of Species Body Size." *Science* 321, no. 5887 (2008): 399–401.

Cooke, Robert S. C., Felix Eigenbrod, and Amanda E. Bates. "Projected losses of global mammal and bird ecological strategies." *Nature Communications* 10, no. 1 (2019). https://doi.org/10.1038/s41467-019-10284-z.

Cools, Annamieke. K. A., Alain. J. M Van Hout, and Mark. H. J. Nelissen. "Canine reconciliation and third-party-initiated postconflict affiliation: do peacemaking social mechanisms in dogs rival those of higher primates?" *Ethology* 114 (2008): 53–63. https://onlinelibrary.wiley.com/doi/abs/10.1111/j.1439-0310.2007.01443.x.

Corrieri, Luca, Marco Adda, Ádám Miklósi, and Enikő Kubinyi. "Companion and free-ranging Bali dogs: Environmental links with personality traits in an endemic dog population of South East Asia." *PLOS ONE* 13, no. 6 2018): e0197354. https://doi.org/10.1371/journal.pone.0197354.

Dagg, Anne Innis. *The Social Behavior of Older Animals*. Baltimore: Johns Hopkins, 2009.

Dale, Rachel, Sylvain Palma-Jacinto, Sarah Marshall-Pescini, and Friederike Range. "Wolves, but not dogs, are prosocial in a touch screen task." *PLOS ONE* 14, no. 5 (2019): e0215444. https://doi.org/10.1371/journal.pone.0215444.

Dale, Rachel, Friederike Range, Laura Stott, Kurt Kotrschal, and Sarah Marshall-Pescini. "The influence of social relationship on food tolerance in wolves and dogs." *Behavioral Ecology and Sociobiology* 71 (2017): 107. https://doi.org/10.1007/s00265-017-2339-8.

Daniela, Sarah E., Rachel E. Fanelli, Amy Gilbert, and Sarah Benson-Amram. "Behavioral flexibility of a generalist carnivore." *Animal Cognition* 22, no. 3 (2019): 387–96. https://doi.org/10.1007/s10071-019-01252-7.

Daniels, Thomas. "Conspecific Scavenging by a Young Domestic Dog." *Journal of Mammalogy* 68, no. 2 (1987): 416–18.

———. "The Social Organization of Free Ranging Urban Dogs. II. Estrous Groups and the Mating System." *Applied Animal Ethology* 10 (1983): 365–73.

Daniels, Thomas, and Marc Bekoff. "Domestication, Exploitation, and Rights." In *Explanation, Evolution, and Adaptation*, edited by Marc Bekoff and Dale Jamieson, 345–77. Vol. 2 of *Interpretation and Explanation in the Study of Animal Behavior.* Boulder, CO: Westview Press, 1990.

———. "Spatial and Temporal Resource Use by Feral and Abandoned Dogs." *Ethology* 81 (1989): 300–12.

Daniels, Thomas J., and Marc Bekoff. "Population and Social Biology of Free-Ranging Dogs, *Canis familiaris*." *Journal of Mammalogy* 70 (1989): 754–62.

———. "Feralization: The Making of Wild Domestic Animals." *Behavioural Processes* 19 (1989): 79–94.

Davis, Matt, Søren Faurby, and Jens-Christian Svenning. "Mammal diversity will take millions of years to recover from the current biodiversity crisis." *Proceedings of the National Academy of Sciences* 115, no. 44 (2018): 11262–67. https://doi.org/10.1073/pnas.1804906115.

Delon, Nicolas. "Pervasive captivity and urban wildlife." *Ethics, Policy and Environment* 23, no. 2 (2020): 123–43. https://doi.org/10.1080/21550085.2020.1848173.

Derr, Mark. *Dog's Best Friend: Annals of the Dog-Human Relationship.* Chicago: University of Chicago Press, 2004.

———. *A Dog's History of America: How Our Best Friend Explored, Conquered, and Settled a Continent.* Albany, CA: North Point Press, 2004.

———. *How the Dog Became the Dog: From Wolves to Our Best Friends.* New York: Abrams Press, 2011.

———. "Shifting Perspectives on How Dogs Came to Be Dogs." *Psychology Today*, September 23, 2019. Accessed April 15, 2020. https://www.psychologytoday.com/us/blog/dogs-best-friend/201909/shifting-perspectives-how-dogs-came-be-dogs.

Diamond, Jared. *Upheaval: How Nations Cope with Crisis and Change.* New York: Penguin Books, 2019.

Doherty, Tim S., Chris R. Dickman, Alistair S. Glen, Thomas M. Newsome, Dale G. Nimmo, Euan G. Ritchie, Abi T. Vanak, and Aaron J. Wirsing. "The global impacts of domestic dogs on threatened vertebrates." *Biological Conservation* 210 (2017): 56–59.

Donfrancesco, Valerio, Paolo Ciucci, Valeria Salvatori, David Benson, Liselotte Wesley Andersen, Elena Bassi, Juan Carlos Blanco, Luigi Boitani, Romolo Caniglia, Antonio Canu, et al. "Unravelling the Scientific Debate on How to Address Wolf-Dog Hybridization in Europe." *Frontiers in Ecology and Evolution* 7 (2019). https://doi.org/10.3389/fevo.2019.00175.

Doomsday Preppers. Produced by Sharp Entertainment NGC Studios/Dominique Andrews Brian Stone. Aired February 7, 2012–August 28, 2014, on National Geographic channel. Original release. https://www.nationalgeographic.com.au/tv/doomsday-preppers/.

Drea, Christine, and Alissa Carter. "Cooperative problem solving in a social carnivore." *Animal Behaviour* 78 (2009): 967–77.

Edmunds, Grace L., Matthew J. Smalley, Sam Beck, Rachel J. Errington, Sara Gould, Helen Winter, Dave C. Brodbelt, Dan G. O'Neill. "Dog breeds and body conformations with predisposition to osteosarcoma in the UK: a case-control study." *Canine Medicine and Genetics* 8 (2021). https://doi.org/10.1186/s40575-021-00100-7.

Faragó, Tamás, Peter Pongrácz, Ádám Miklósi, Ludwig Huber, Zsófia Virányi, Friederike Range. "Dogs' Expectation about Signalers' Body Size by Virtue of Their Growls." *PLOS ONE* 5, no. 12 (2010): e15175. https://doi.org/10.1371/journal.pone.0015175.

Fawcett, Anne, Vanessa Barrs, Magdoline Awad, Georgina Child, Laurencie Brunel, Erin Mooney, Fernando Martinez-Taboada, Beth McDonald, and Paul McGreevy. "Consequences and Management of Canine Brachycephaly in Veterinary Practice: Perspectives from Australian Veterinarians and Veterinary Specialists." *Animals* (2018). https://www.mdpi.com/2076-2615/9/1/3/htm.

Feddersen-Petersen, Dorit. "Social Behaviour of Dogs and Related Canids." In *The Behavioural Biology of Dogs*, edited by Per Jensen, 105–19. Oxfordshire, UK: CAB International, 2007.

Font, Enrique. "Spacing and social organization: urban stray dogs revisited." *Applied Animal Behaviour Science* 17 (1987): 319–28. https://doi.org/10.1016/0168-1591(87)90155-9.

Fox, Michael W. *Behaviour of Wolves, Dogs, and Related Canids*. New York: Harper & Row, 1972.

———, ed. *The Wild Canids: Their Systematics, Behavioral Ecology and Evolution*. New York: Litton, 1975 (reprinted 2009 by Dogwise Publishing).

Francis, Richard C. *Domesticated: Evolution in a Man-Made World*. New York: W. W. Norton, 2016.

Frantz, Laurent A. F., Victoria E. Mullin, Maud Pionnier-Capitan, Ophélie Lebrasseur, Morgane Ollivier, Angela Perri, Anna Linderholm, Valeria Mattiangeli, Matthew D. Teasdale, Evangelos A. Dimopoulos, et al. "Genomic and archaeological evidence suggest a dual origin of domestic dogs." *Science* 352, no. 6290 (2016): 1228–31. https://doi.org/10.1126/science.aaf3161.

Fredrickson, Richard J., and Philip W. Hedrick. "Dynamics of hybridization and introgression in red wolves and coyotes." *Conservation Biology* 20 (2006): 1272–83.

Freedman, Adam H., Ilan Gronau, Rena M. Schweizer, Diego Ortega-Del Vecchyo, Eunjung Han, Pedro M. Silva, Marco Galaverni, Zhenxin Fan, Peter Marx, Belen Lorente-Galdos, et al. "Genome Sequencing Highlights the Dynamic Early History of Dogs." *PLOS Genetics* 10, no. 1 (2014): e1004016. https://doi.org/10.1371/journal.pgen.1004016.

Freedman, Daniel G., John A. King, and Oliver Elliot. "Critical Period in the Social Development of Dogs." *Science* 133 (1961): 1016–17. https://doi.org/10.1126/science.133.3457.1016. PMID: 13701603.

Galis, Frietson, Inke Van der Sluijs, Tom J. M. V. Van Dooren, Johan A. J. Metz, Marc Nussbaumer. "Do large dogs die young?" *Journal of Experimental Zoology Part B: Molecular and Developmental Evolution* 308, no. 2 (2007): 119–26.

Galov, Ana, Elena Fabbri, Romolo Caniglia, Haidi Arbanasić, Silvana Lapalombella, Tihomir Florijančić, Ivica Bošković, Marco Galaverni, and Ettore Randi. "First evidence of hybridization between golden jackal (*Canis aureus*) and domestic dog (*Canis familiaris*) as revealed by genetic markers." *Royal Society Open Science* 2, no. 12 (2015): 150450. https://royalsocietypublishing.org/doi/10.1098/rsos.150450.

Gamborg, Christian, Bart Gremmen, Stine B. Christiansen, and Peter Sandøe. "De-Domestication: Ethics at the Intersection of Landscape Restoration and Animal Welfare." *Environmental Values* 19, no. 1 (2010): 57–78. https://doi.org/10.3197/096327110X485383.

Gardner, Howard. *Frames of Mind: The Theory of Multiple Intelligences.* New York: Basic Books, 1983.

Geffen, Eli, Michael Kam, Reuven Hefner, Pall Hersteinsson, Anders Angerbjörn, Love Dalèn, Eva Fuglei, Karin Norèn, Jennifer. R. Adams, John Vucetich, et al. "Kin encounter rate and inbreeding avoidance in canids." *Molecular Ecology* (2011): 5348–56. https://www.ncbi.nlm.nih.gov/pubmed/22077191.

Ghosh, B., D. K. Choudhuri, and B. Pal. "Some aspects of the sexual behaviour of stray dogs, *Canis familiaris*." *Applied Animal Behaviour Science* 13, nos. 1–2 (1984): 113–27.

Gibson, Johanna. *Owned, An Ethological Jurisprudence of Property: From the Cave to the Commons.* Milton Park, Abingdon-on-Thames, Oxfordshire, UK: Routledge, 2020.

Girman, Derek. J., M. G. L. Mills, Eli Geffen, and Robert. K. Wayne. "A molecular genetic analysis of social structure, dispersal, and interpack relationships of the African wild dog (*Lycaon pictus*)." *Behavioral Ecology and Sociobiology* 40 (1997): 187–98.

Gomes da Silva, Roberto, and Alex Sandro Campos Maia. *Principles of Animal Biometeorology.* Dordrecht, Netherlands: Springer, 2013.

Gompert, Zachariah, and C. Alex Buerkle. "What, if anything, are hybrids: enduring truths and challenges associated with population structure and gene flow." *Evolutionary Applications* 9 (2016): 909–23. https://doi.org/10.1111/eva.12380.

Gompper, Matthew E. "The dog–human–wildlife interface: assessing the scope of the problem." In *Free-Ranging Dogs and Wildlife Conservation*, edited by Matthew E. Gompper, 9–54. Oxford: Oxford University Press, 2014.

———. *Free-Ranging Dogs and Wildlife Conservation*. Oxford: Oxford University Press, 2014.

Goodwin, Deborah, John W. S. Bradshaw, and Stephen M. Wickens. "Paedomorphosis affects agonistic visual signals of domestic dogs." *Animal Behaviour* 53 (1997): 297–304.

Griffin, Donald. *Animal Minds*. Chicago: University of Chicago, Press. 1992.

Haraway, Donna J. *When Species Meet*. Minneapolis: University of Minnesota Press, 2008.

Harrison, Richard G., and Erica L. Larson. "Hybridization, Introgression, and the Nature of Species Boundaries." *Journal of Heredity* 105 (2014): 795–809.

Healy, Kevin, Thomas H. G. Ezard, Owen R. Jones, Roberto Salguero-Gómez, and Yvonne M. Buckley. "Animal life history is shaped by the pace of life and the distribution of age-specific mortality and reproduction." *Nature Ecology and Evolution* 3 (2019): 1217–24. https://doi.org/10.1038/s41559-019-0938-7.

Hecht, Erin E., Jeroen B. Smaers, William J. Dunn, Marc Kent, Todd M. Preuss, and David A. Gutman. "Significant neuroanatomical variation among domestic dog breeds." *Journal of Neuroscience* 2 (2019): 303–19. https://doi.org/10.1523/JNEUROSCI.0303-19.2019.

Heid, Markham. "How Dogs Would Fare without Us," *Time* special issue, "How Dogs Think" (2018): 60–65.

Hemmer, Helmut. *Domestication: The Decline of Environmental Appreciation*. Translated by Neil Beckhaus. Cambridge: Cambridge University Press, 1990.

Henry, J. David. *Red Fox: The Catlike Canine*. Washington, DC: Smithsonian Institution Press, 1986.

Heppenheimer, Elizabeth, Kristin E. Brzeski, Ron Wooten, William Waddell, Linda Y. Rutledge, Michael J. Chamberlain, Daniel R. Stahler, Joseph W. Hinton, and Bridgett M. VonHoldt. "Rediscovery of Red Wolf Ghost Alleles in a Canid Population along the American Gulf Coast." *Genes* 9, no. 12 (2018). https://doi.org/10.3390/genes9120618.

Herborn, Katherine A., Ross MacLeod, Will T. S. Miles, Anneka N. B. Schofield, Lucile Alexander, and Kathryn E. Arnold. "Personality in captivity reflects personality in the wild." *Animal Behaviour* 79 (2010): 835–43.

Hernandez-Avalos, Ismael Daniel Mota-Rojas, Patricia Mora-Medina, Julio Martínez-Burnes, Alejandro Casas Alvarado, Antonio Verduzco-Mendoza, Karina Lezama-García, and Adriana Olmos-Hernandez. "Review of different methods used for clinical recognition and assessment of pain in dogs and cats." *International Journal of Veterinary Science and Medicine* 7, no. 1 (2019): 43–54.

Herzog, Hal. "Is a Love of Dogs Mostly a Matter of Where You Live? Global dog demographics show the impact of culture on human-pet relationships." *Psychology Today*. Accessed April 14, 2020. https://www.psychologytoday.com/us/blog/animals-and-us/201908/is-love-dogs-mostly-matter-where-you-live.

Hiby, Elly F., Nicola J. Rooney, and John W. S. Bradshaw. "Behavioural and physiological responses of dogs entering re-homing kennels." *Physiology and Behavior* 89 (2006): 385–91.

Høgåsen, H. R, C. Er, A. Di Nardo, and P. Dalla Villa. "Free-roaming dog populations: A cost-benefit model for different management options, applied to Abruzzo, Italy." *Preventive Veterinary Medicine* 112, nos. 3–4 (2013): 401–13. https://doi.org/10.1016/j.prevetmed.2013.07.010. PMID: 23973012.

Holekamp, Kay E., and Sarah Benson-Amram. "The evolution of intelligence in mammalian carnivores." *Interface Focus* 7 (2017): 20160108.

Horowitz, Alexandra, ed. *Domestic Dog Cognition and Behavior: The Scientific Study of Canis Familiaris*. New York: Springer 2014.

Horschler, Daniel J., Brian Hare, Josep Call, Juliane Kaminski, Ádám Miklósi, and Evan L MacLean. "Absolute brain size predicts dog breed differences in executive function." *Animal Cognition* 22 (2019): 187–98. https://doi.org/10.1007/s10071-018-01234-1.

Horváth, Zsuzsánna, Igyártó Botond-Zoltán, Attila Magyar, and Ádám Miklósi. "Three different coping styles in police dogs exposed to a short-term challenge." *Hormones and Behavior* 52, no. 5 (2007): 621–30.

Hunter, Luke. *Carnivores of the World.* Princeton: Princeton University Press, 2018.

Inoue, Mai, A. Hasegawa, Y. Hosoi, and K. Sugiura. "A current life table and causes of death for insured dogs in Japan." *Preventive Veterinary Medicine* 120, no. 2 (2015): 210–18.

Jensen, Per. *The Behavioural Biology of Dogs.* Oxfordshire, UK: CAB International, 2007.

———. "Mechanisms and Function in Behaviour." In *The Behavioural Biology of Dogs*, edited by Per Jensen, 61–75. Oxfordshire, UK: CAB International, 2007.

Jensen, Per, Mia Persson, Dominic Weight, Martin Johnsson, Ann-Sofie Sundman, and Lina Roth. "The Genetics of How Dogs Became Our Social Allies." *Current Directions in Psychological Science* 25, no. 5 (2016): 334–38.

Johnson-Ulrich, Lily, Sarah Benson-Amram, and Kay E. Holekamp. "Fitness Consequences of Innovation in Spotted Hyenas." *Frontiers in Ecology and Evolution* 22 (2019). https://doi.org/10.3389/fevo.2019.00443.

Johnston, Angie M., Courtney Turrin, Lyn Watson, Alyssa M. Arre, and Laurie R. Santos. "Uncovering the origins of dog–human eye contact: dingoes establish eye contact more than wolves, but less than dogs." *Animal Behaviour* 133 (2017): 123–29.

Jones, Amanda, and Samuel D. Gosling. "Temperament and personality in dogs (*Canis familiaris*): A review and evaluation of past research." *Applied Animal Behaviour Science* 95 (2005): 1–53.

Jung, Christoph, and Daniela Pörtl. "Scavenging Hypothesis: Lack of Evidence for Dog Domestication on the Waste Dump." *Dog Behavior* 2 (2018): 41–56.

Kaminski, Juliane, and Sarah Marshall-Pescini, *The Social Dog: Behaviour and Cognition.* San Diego, CA: Elsevier, 2014.

Kaminski, Juliane, Bridget M. Waller, Rui Diogo, Adam Hartstone-Rose, and Anne M. Burrows. "Evolution of facial muscle anatomy in dogs." *Proceedings of the National Academy of Sciences* 116, no. 29 (2019) 14677–81. https://doi.org/10.1073/pnas.1820653116.

Kean, Hilda. *The Great Dog and Cat Massacre: The Real Story of World War Two's Unknown Tragedy*. Chicago: University of Chicago Press, 2017.

Kitchtenham, Kate. *Streunerhunde: Von Moskaus U-Bahn-Hunden bis Indiens Underdogs* (German Edition). Stuttgart: Franckh-Kosmos, 2020.

Kjelgaard, Jim. *Desert Dog*. New York: Bantam Skylark, 1979.

Koolhaas, J. M., S. M. Korte, S. F. De Boer, B. J. Van Der Vegt, C. G. Van Reenen, H. Hopster, I. C. De Jong, M. A. W. Ruis, and H. J. Blokhuis. "Coping styles in animals: current status in behavior and stress-physiology." *Neuroscience and Biobehavioral Reviews* 23, no. 7 (1999): 925–35.

Kopaliani, Natia, Maia Shakarashvili, Zurab Gurielidze, Tamar Qurkhuli, and David Tarkhnishvili. "Gene flow between wolf and shepherd dog populations in Georgia (Caucasus)." *Journal of Heredity* 105, no. 3 (2014): 345–53.

Kraus, Cornelia, Samuel Pavard, and Daniel E. L. Promislow, "The Size–Life Span Trade-Off Decomposed: Why Large Dogs Die Young." *American Naturalist* 181, no. 4 (April 2013): 492–505.

Kronfeld-Schor, Noga, Guy Bloch, and William J. Schwartz. "Animal clocks: when science meets nature." *Proceedings. Biological sciences* 280, no. 1765 (2013): 20131354. https://doi.org/10.1098/rspb.2013.1354.

Lark, Karl G., Kevin Chase, and Nathan B. Sutter. "Genetic architecture of the dog: sexual size dimorphism and functional morphology." *Trends in Genetics* 22, no. 10 (2006): 537–44. https://doi.org/10.1016/j.tig.2006.08.009.

Larson, Rachel N., Justin L. Brown, Tim Karels, Seth P. D. Riley. "Effects of urbanization on resource use and individual specialization in coyotes (*Canis latrans*) in southern California." *PLOS ONE* 15, no. 2 (2020): e0228881. https://doi.org/10.1371/journal.pone.0228881.

Lazzaroni, Martina, Friederike Range, Jessica Backes, Katrin Portele, Katharina Scheck, Sarah Marshall-Pescini. "The Effect of Domestication and Experience on the Social Interaction of Dogs and Wolves With a Human Companion." *Frontiers in Psychology*, 11 (2020): 785. https://doi.org/10.3389/fpsyg.2020.00785.

Lazzaroni, Martina, Friederike Range, Lara Bernasconi, Larissa Darc, Maria Holtsch, Roberta Massimei, Akshay Rao, and Sarah Marshall-Pescini. "The role of life experience in affecting persistence: A comparative study between free-ranging dogs, pet dogs and captive pack dogs." *PLOS ONE* 14, no. 4 (2019): e0214806. https://doi.org/10.1371/journal.pone.0214806.

Lemaître, Jean-François, Victor Ronget, Morgane Tidière, Dominique Allainé, Vérane Berger, Aurélie Cohas, Fernando Colchero, Dalia A. Conde, Michael Garratt, András Liker, Gabriel A. Marais, Alexander Scheuerlein, Tamás Székely, and Jean-Michel Gaillard. "Sex differences in adult lifespan and aging rates of mortality across wild mammals." *Proceedings of the National Academy of Sciences* (2020): 201911999. https://doi.org/10.1073/pnas.1911999117.

Lescureux, Nicolas, and John D. C. Linnell. "Warring brothers: The complex interactions between wolves (*Canis lupus*) and dogs (*Canis familiaris*) in a conservation context." *Biological Conservation* 171 (2014): 232–45. https://doi.org/10.1016/j.biocon.2014.01.032.

Leyhausen, Paul. "The Communal Organization of Solitary Animals." *Symposia of the Zoological Society of London* 14 (1965): 249–62.

Li, Yan, Bridgett M. Vonholdt, Andy Reynolds, Adam R. Boyko, Robert K. Wayne, Dong-Dong Wu, and Ya-Ping Zhang, "Artificial Selection on Brain-Expressed Genes during the Domestication of Dog." *Molecular Biology and Evolution* 30, no. 8 (2013): 1867–76. https://doi.org/10.1093/molbev/mst088.

Lockyear, K. M., W. T. Waddell, K. L. Goodrowe, and S. E. MacDonald. "Retrospective Investigation of Captive Red Wolf Reproductive Success in Relation to Age and Inbreeding." *Zoo Biology* 28 (2009): 214–29.

Lord, Kathryn, Mark Feinstein, and Bradley Smith, Raymond Coppinger. "Variation in reproductive traits of members of the genus *Canis* with special attention to the domestic dog (*Canis familiaris*)." *Behavioral Processes* 92 (2013): 131–42. https://doi.org/10.1016/j.beproc.2012.10.009.

Lorenz, Konrad. *Man Meets Dog.* New York: Routledge, 2002.

Lorimer, Jamie. *Wildlife in the Anthropocene: Conservation after Nature.* Minneapolis: University of Minnesota Press, 2015.

Losos, Jonathan. *Improbable Destinies: Fate, Chance, and the Future of Evolution.* New York: Riverhead Books, 2017.

MacArthur, Robert H., and Edward O. Wilson. *The Theory of Island Biogeography.* Princeton: Princeton University Press, 1967.

Macdonald, D. W., and G M. Carr. "Variation in dog society: between resource dispersion and social flux." In *The Domestic Dog*, 2nd ed., edited by James Serpell, 319–41. Cambridge: Cambridge University Press, 2017.

Macdonald, David, Scott Creel, and Michael G. H. Mills. "Canid Society." In *Biology and Conservation of Wild Canids*, edited by David W. Macdonald and Claudio Sillero-Zubiri, 85–106. New York: Oxford University Press, 2004.

Macdonald, David W., and Claudio Sillero-Zubili, eds. *The Biology and Conservation of Wild Canids.* New York: Oxford University Press, 2004.

Macdonald, David. W., Liz. A. D. Campbell, Jan. F. Kamler, Jorgelina Marino, Geraldine Werhahn, and Claudio Sillero-Zubiri. "Monogamy: Cause, Consequence, or Corollary of Success in Wild Canids?" *Frontiers in Ecology and Evolution* 7 (2019): 341. https://doi.org/10.3389/fevo.2019.00341.

Maglieri, Veronica, Filippo Bigozzi, Marco Germain Riccobono, and Elisabetta Palagi. "Levelling playing field: synchronization and rapid facial mimicry in dog-horse play." *Behavioural Processes* 174 (2020): 104104. https://doi.org/10.1016/j.beproc.2020.104104.

Majumder, Sreejani Sen, and Anindita Bhadra. "When Love Is in the Air: Understanding Why Dogs Tend to Mate When It Rains." *PLOS ONE* 10, no. 12 (2015). https://journals.plos.org/plosone/article?id=10.1371/journal.pone.0143501.

Majumder, Sreejani Sen, Paul Manabi, Sau Shubhra, and Anindita Bhadra. "Denning habits of free-ranging dogs reveal preference for human proximity." *Scientific Reports* 6 (2016): 32014.

Marshall-Pescini, Sarah, Franka S. Schaebs, Alina Gaugg, Anne Meinert, Tobias Deschner, and Friederike Range. "The Role of Oxytocin in the Dog–Owner Relationship." *Animals* 9 (2019): 792. https://www.mdpi.com/2076-2615/9/10/792.

Marshall-Pescini, Sarah, Jonas F. L. Schwarz, Inga Kostelnik, Zsófia Virányi, and Friederike Range. "Importance of a species' socioecology: Wolves outperform dogs in a conspecific cooperation task." *Proceedings of the National Academy of Sciences* 114 (2017): 11793–98. https://www.pnas.org/content/114/44/11793.

Marshall-Pescini, Sarah, Paola Valsecchi, Irena Petak, Pier. Attilio Accorsi, Emanuela. P. Previde. "Does training make you smarter? The effects of training on dogs' performance (*Canis familiaris*) in a problem solving task." *Behavioural Processes* 78, no. 3 (2008): 449–54. PMID: 18434043.

Matter, Hans, and Thomas Daniels, "Dog Ecology and Population Biology." In *Dogs, Zoonoses, and Public Health*, edited by Calum Macpherson, François Meslin, and Alexander Wandeler, 17–62. New York: CABI Publishing, 2000.

McCain, Christy M., and Sarah R. King. "Body size and activity times mediate mammalian responses to climate change." *Global Change Biology* 20, no. 6 (2014): 1760–69. https://doi.org/10.1111/gcb.12499.

McGreevy, Paul, Tanya D. Grassi, and Alison M. Harman. "A strong correlation exists between the distribution of retinal ganglion cells and nose length in the dog." *Brain and Behavioral Science* 63, no. 1 (2004): 13–22.

McIntyre, Rick. *The Rise of Wolf 8: Witnessing the Triumph of Yellowstone's Underdog.* New York: Greystone, 2019.

McIntyre, R., J. B. Theberge, M. T. Theberge, D. W. Smith. "Behavioral and ecological implications of seasonal variation in the frequency of daytime howling by Yellowstone wolves." *Journal of Mammalogy* 98, no. 3 (2017): 827–34. https://doi.org/10.1093/jmammal/gyx034.

Mech, David. "Disproportionate Sex Ratios of Wolf Pups." *Journal of Wildlife Management* 39, no. 4 (1975): 737–40.

Mech, L. David. *The Wolf: The Ecology and Behavior of an Endangered Species.* Minneapolis: University of Minnesota Press, 1981.

Mech, L. David, and Luigi Boitani, eds. *Wolves: Behavior, Ecology, and Conservation.* University of Chicago Press: Chicago, 2003.

Miklósi, Ádám. *Dog Behaviour: Evolution and Cognition*, 2nd ed. New York: Oxford University Press, 2016.

———. "Human-Animal Interactions and Social Cognition in Dogs." In *The Behavioural Biology of Dogs*, edited by Per Jensen, 207–22. Oxfordshire, UK: CAB International, 2007.

Miklósi, Ádám, Enikö Kubinyi, József Topál, Márta Gácsi, ZsófiaVirányi, and Vilmos Csány. "A Simple Reason for a Big Difference: Wolves Do Not Look Back at Humans, but Dogs Do." *Current Biology* 13, no. 9 (2003): 763–66.

Miternique, Capellà Hugo, and Florence Gaunet. "Coexistence of Diversified Dog Socialities and Territorialities in the City of Concepción, Chile." *Animals* 10 (2020): 298.

Morey, Darcy F. *Dogs: Domestication and the Development of a Social Bond.* Cambridge: Cambridge University Press, 2010.

Mugford, Roger. "Behavioural Disorders." In *The Behavioural Biology of Dogs*, edited by Per Jensen, 225–42. Oxfordshire, UK: CAB International, 2007.

Müller, Corsin A., Christina Mayer, Sebastian Dörrenberg, Ludwig Huber, and Friederike Range. "Female but not male dogs respond to a size constancy violation." *Biology Letters* 7, no. 5 (2011): 689–91. https://doi.org/10.1098/rsbl.2011.0287.

Nagasawa, Miho, Shouhei Mitsui, Shiori En, Nobuyo Ohtani, Mitsuaki Ohta, Yasuo Sakuma, Tatsushi Onaka, Kazutaka Mogi, Takefumi Kikusui. "Social evolution. Oxytocin-gaze positive loop and the coevolution of human-dog bonds." *Science* 348, no. 6232 (2015): 333–36. https://doi.org/10.1126/science.1261022.

Nesbitt, William H. "Ecology of a Feral Dog Pack on a Wildlife Refuge." In *The Wild Canids: Their Systematics, Behavioral Ecology and Evolution*, edited by Michael W. Fox, 391–96. New York: Litton, 1975.

Nicholas, Frank W., Elizabeth R. Arnott, Paul D. McGreevy. "Hybrid vigour in dogs?" *Veterinary Journal* 214 (2016): 77–83.

Packard, Jane M., Ulysses S. Seal, L. David Mech, and Edward D. Plotka. "Causes of Reproductive Failure in Two Family Groups of Wolves

(*Canis lupus*)." *Zeitschrift für Tierpsychologie* 68 (1985): 24–40. https://doi.org/10.1111/j.1439-0310.1985.tb00112.x.

Packer, Rowena M. A., Anke Hendricks, Michael. S. Tivers, Charlotte. C. Burn. "Impact of Facial Conformation on Canine Health: Brachycephalic Obstructive Airway Syndrome." *PLOS ONE* 10, no. 10 (2015): e0137496. https://doi.org/10.1371/journal.pone.0137496.

Packer, Rowena. M. A., Anke Hendricks, Holger. A. Volk, Nadia. K. Shihab, Charlotte. C. Burn. "How Long and Low Can You Go? Effect of Conformation on the Risk of Thoracolumbar Intervertebral Disc Extrusion in Domestic Dogs." *PLOS ONE* 8, no. 7 (2013): e69650. https://doi.org/10.1371/journal.pone.0069650.

Pal, Sunil Kumar. "Factors influencing intergroup agonistic behaviour in free-ranging domestic dogs (*Canis familiaris*)." *Acta Ethologica* 18 (2015): 209–20.

———. "Mating system of free-ranging dogs (*Canis familiaris*)." *International Journal of Zoology* (2011): 1–10.

———. "Parental care in free-ranging dogs, *Canis familiaris*." *Applied Animal Behaviour Science* 90, no. 1 (2005): 31–47.

———. "Play behaviour during early ontogeny in free-ranging dogs (*Canis familiaris*)." *Applied Animal Behaviour Science* 126, nos. 3–4 (2010): 140–53.

———. "Population ecology of free-ranging urban dogs in West Bengal, India." *Acta Theriologica* 46, no. 2 (2001): 69–78.

Pal, S. K., B. Ghosh, S. Roy. "Agonistic behaviour of free-ranging dogs (*Canis familiaris*) in relation to season, sex and age." *Applied Animal Behaviour Science* 59, no. 4 (1998): 331–48.

———. "Inter- and intra-sexual behaviour of free-ranging dogs (*Canis familiaris*)." *Applied Animal Behaviour Science* 62 (1999): 267–78.

Pal, S. K., S. Roy, and B. Ghosh. "Pup rearing: the role of mothers and allomothers in free-ranging domestic dogs." *Applied Animal Behaviour Science* 234 (2020). https://doi.org/10.1016/j.applanim.2020.105181.

Palagi, Elisabetta, and Giada Cordoni. "Postconflict third-party affiliation in *Canis lupus*: do wolves share similarities with the great apes?"

Animal Behaviour 78 (2009): 979–86. https://doi.org/10.1016/j.anbehav.2009.07.017.

Palagi, Elisabetta, Velia Nicotra, and Giada Cordoni. "Rapid mimicry and emotional contagion in domestic dogs." *Royal Society Open Science* 2, no. 12 (2015): 150505. https://doi.org/10.1098/rsos.150505.

Parker, Heidi G., Alexander Harris, Dayna L. Dreger, Brian W. Davis, and Elaine A. Ostrander. "The bald and the beautiful: hairlessness in domestic dog breeds." *Philosophical transactions of the Royal Society of London. Series B, Biological sciences* 372, no. 1713 (2017): 20150488. https://doi.org/10.1098/rstb.2015.0488.

Paschoal, Ana Maria, Rodrigo L. Massara, Larissa Bailey, Paul F. Doherty Jr., Paloma M. Santos, Adriano Paglia, Andre Hirsch, and Adriano G. Chiarello. "Anthropogenic Disturbances Drive Domestic Dog Use of Atlantic Forest Protected Areas." *Tropical Conservation Science* 11 (2018): 1–14.

Paul, Manabi, and Anindita Bhadra. "The great Indian joint families of free-ranging dogs." *PLOS ONE* 13, no. 5 (2018). https://journals.plos.org/plosone/article?id=10.1371/journal.pone.0197328.

Paul, Manabi, Sreejani Sen Majumder, Shubhra Sau, Anjan K. Nandi, and Anindita Bhadra. "High early life mortality in free-ranging dogs is largely influenced by humans." *Scientific Reports* 6 (2016): 19641. https://doi.org/10.1038/srep19641.

Pérez-Manrique, Ana, and Antoni Gomila. "The comparative study of empathy: sympathetic concern and empathic perspective-taking in non-human animals." *Biological Reviews* 93 (2018): 248–69.

Perry, Laura R., Bernard MacLennan, Rebecca Korven, and Timothy A. Rawlings. "Epidemiological study of dogs with otitis externa in Cape Breton, Nova Scotia." *Canadian Veterinary Journal/La Revue vétérinaire canadienne* 58, no. 2 (2017): 168–74.

"Pets by the Numbers," HumanePro. Accessed April 15, 2020. https://humanepro.org/page/pets-by-the-numbers.

Pierce, Jessica. "Beyond Humans: Dog Utopia or Dog Dystopia?" *Psychology Today* (blog), October 18, 2018. https://www.psychologytoday.com/ca/blog/all-dogs-go-heaven/201810/beyond-humans-dog-utopia-or-dog-dystopia.

———. *Run, Spot, Run: The Ethics of Keeping Pets*. Chicago: University of Chicago Press, 2016.

Pongrácz, Péter, and Sára S. Sztruhala. "Forgotten, But Not Lost—Alloparental Behavior and Pup–Adult Interactions in Companion Dogs." *Animals* 9 (2019): 1011.

Price, E. O. "Behavioral development in animals undergoing domestication." *Applied Animal Behaviour Science* 65 (1999): 245–71.

Price, Edward O. "Behavioral Aspects of Animal Domestication." *Quarterly Review of Biology* 59, no. 1 (1984): 1–26.

Purcell, Brad. *Dingo*. Collingwood, Australia: CSIRO Publishing, 2010.

Quervel-Chaumette, Mylene, Viola Faerber, Tamás Faragó Sarah Marshall-Pescini, Friederike Range. "Investigating Empathy-Like Responding to Conspecifics' Distress in Pet Dogs." *PLOS ONE* 11, no. 4 (2016): e0152920. https://doi.org/10.1371/journal.pone.0152920.

Ralls, Katherine, P H. Harvey, and A. M. Lyles, and M. Soulé. "Inbreeding in natural populations of birds and mammals." In *Conservation Biology: The Science of Scarcity and Diversity*, edited by Michael E. Soulé, 35–56. Sunderland, MA: Sinauer Associates, 1986.

Range, Friederike, Alexandra Kassis, Michael Taborsky, Mónica Boada, and Sarah Marshall-Pescini. "Wolves and dogs recruit human partners in the cooperative string-pulling task." *Scientific Reports* 9 (2019): 17591. https://doi.org/10.1038/s41598-019-53632-1.

Range, Friederike, Sarah Marshall-Pescini, Corrina Kratz, and Zsófia Virányi. "Wolves lead and dogs follow, but they both cooperate with humans." *Scientific Reports* 9 (2019): 3796.

Riach, Anna C., Rachel Asquith, and Melissa L. D. Fallon. "Length of time domestic dogs (*Canis familiaris*) spend smelling urine of gonadectomised and intact conspecifics." *Behavioural Processes* 142 (2017): 138–40. https://pubmed.ncbi.nlm.nih.gov/28689817/.

Ritchie, Euan G., Christopher R. Dickman, Mike Letnic, and Abi Tamim Vanak. "Dogs as predators and trophic regulators." In *Free-Ranging Dogs and Wildlife Conservation*, edited by Matthew Gompper, 55–68. New York: Oxford University Press, 2014.

Robinson, Jacqueline A., Jannikke Räikkönen, Leah M. Vucetich, John A. Vucetich, Rolf O. Peterson, Kirk E. Lohmueller, and Robert K. Wayne.

"Genomic signatures of extensive inbreeding in Isle Royale wolves, a population on the threshold of extinction." *Science Advances* 5, no. 5 (2019).

Rueb. Emily S., and Niraj Chokshi. "Labradoodle Creator Says the Breed Is His Life's Regret," *New York Times*, September 25, 2019. Accessed April 15, 2020. https://www.nytimes.com/2019/09/25/us/labradoodle-creator-regret.html.

Saetre, Peter, Julia Lindberg, Jennifer A. Leonard, Kerstin Olsson, Ulf Petterssond, Hans Ellegrena, Tomas F. Bergstro, Carles Vila, and Elena Jazin. "From wild wolf to domestic dog: gene expression changes in the brain." *Molecular Brain Research* 126 (2004): 198–206.

Salt, Carina, Penelope J. Morris, Derek Wilson, Elizabeth. M. Lund, and Alexander J. German. "Association between life span and body condition in neutered client-owned dogs." *Journal of Veterinary Internal Medicine* 33, no. 1 (2019): 89–99. https://doi.org/10.1111/jvim.15367.

Samuel, Lydia, Charlotte Arnesen, Andreas. Zedrosser, Frank Rosell. "Fears from the past? The innate ability of dogs to detect predator scents." *Animal Cognition* 23 (2020): 721–29. https://doi.org/10.1007/s10071-020-01379-y.

Sands, Jennifer, and Scott Creel. "Social dominance, aggression and faecal glucocorticoid levels in a wild population of wolves, *Canis lupus*." *Animal Behaviour* 67 (2004): 387–96.

Santicchia, Francesca, Claudia Romeo, Nicola Ferrari, Erik Matthysen, Laure Vanlauwed, Lucas A. Wauters, and Adriano Martinoli. "The price of being bold? Relationship between personality and endoparasitic infection in a tree squirrel." *Mammalian Biology* 97 (2019): 1–8. https://doi.org/10.1016/j.mambio.2019.04.007.

Sarkar, Rohan, Shubhra Sau, and Anindita Bhadra. "Scavengers can be choosers: A study on food preference in free-ranging dogs." *Applied Animal Behaviour Science* 216 (2019): 38–44.

Savolainen, Peter. "Domestication of Dogs." In *The Behavioural Biology of Dogs*, edited by Per Jensen, 21–37. Oxfordshire, UK: CAB International, 2007.

Scandurra, Anna, Alessandra Alterisio, Anna Di Cosmo, and Biagio D'Aniello. "Behavioral and Perceptual Differences between Sexes in Dogs: An Overview." *Animals* 8, no. 9 (2018): 151. https://doi.org/10.3390/ani8090151.

Schenkel, Rudolf. "Expressions Studies on Wolves." *Behaviour* 1 (1947): 81–129.

http://davemech.org/wolf-news-and-information/schenkels-classic-wolf-behavior-study-available-in-english/.

Schilthuizin, Menno. *Darwin Comes to Town*. New York: Picador, 2018.

Scott, John Paul, and John L. Fuller. *Genetics and the Social Behavior of the Dog*. Chicago: University of Chicago Press, 1965.

Seyle, Hans. *The Stress of Life*. New York: McGraw-Hill, 1978.

Shipman, Pat. *The Invaders: How Humans and Their Dogs Drove Neanderthals to Extinction*. Cambridge, MA: Harvard University Press, 2015.

———. "What the dingo says about dog domestication." *Anatomical Record* 304 (2021): 19–30. https://doi.org/10.1002/ar.24517.

Sidorovich, V. E., V. P. Stolyarov, N. N. Vorobei, N. V. Ivanova, and B. Jędrzejewska. "Litter size, sex ratio, and age structure of gray wolves, Canis lupus, in relation to population fluctuations in northern Belarus." *Canadian Journal of Zoology* 85 (2007): 295–300. https://doi.org/10.1139/Z07-001.

Signore, Anthony V., Ying-Zhong Yang, Quan-Yu Yang, Ga Qin, Hideaki Moriyama, Ri-Li Ge, Jay F. Storz. "Adaptive Changes in Hemoglobin Function in High-Altitude Tibetan Canids Were Derived via Gene Conversion and Introgression." *Molecular Biology and Evolution* 36, no. 10 (2019): 2227–37. https://doi.org/10.1093/molbev/msz097.

Silk, Matthew J., Michael. A. Cant, Simona Cafazzo, Eugenia Natoli, Robbie. A. McDonald. "Elevated aggression is associated with uncertainty in a network of dog dominance interactions." *Proceedings of the Royal Society B* 286 (2019): 20190536. http://dx.doi.org/10.1098/rspb.2019.0536.

Silva, Karine, and Liliana de Sousa. "'*Canis empathicus*'? A proposal on dogs' capacity to empathize with humans." *Biology Letters* (2011). http://doi.org/10.1098/rsbl.2011.0083.

Smith, Bradley, ed. *The Dingo Debate: Origins, Behaviour and Conservation*. Collingwood, Australia: CSIRO Publishing, 2015.

Smith, Deborah, Thomas Meier, Eli Geffen, L. David Mech, John W. Burch, Layne G. Adams, Robert K. Wayne. "Is incest common in gray wolf packs?" *Behavioral Ecology* 8 (1997): 384–91.

Sober, Elliot. *The Nature of Selection*. Chicago: University of Chicago Press, 1993.

Špinka, Marek, Ruth C. Newberry, and Marc Bekoff. "Mammalian play: Training for the unexpected." *Quarterly Review of Biology* 76 (2001): 141–68.

Spotte, Stephen. *Societies of Wolves and Free-ranging Dogs*. Oxford: Oxford University Press, 2012.

Stearns, Stephen C. *The Evolution of Life Histories*. Cambridge: Cambridge University Press, 1992.

———. "Life-History Tactics: A Review of the Ideas." *Quarterly Review of Biology* 51, no. 1 (1976): 3–47.

———. "Trade-offs in life history evolution." *Functional Ecology* 3 (1989): 259–68.

Sumner, Rebecca N., Mathew Tomlinson, Jim Craigon, Gary C. W. England, and Richard G. Lea. "Independent and combined effects of diethylhexyl phthalate and polychlorinated biphenyl 153 on sperm quality in the human and dog." *Scientific Reports* 9, no. 1 (2019). https://doi.org/10.1038/s41598-019-39913-9.

Swanson, Heather Anne, Marianne Elisabeth Lien, and Gro B. Ween, eds. *Domestication Gone Wild: Politics and Practices of Multispecies Relations*. Durham, NC: Duke University Press, 2018.

Szabo Birgit, Isabel Damas-Moreira, and Martin J. Whiting. "Can Cognitive Ability Give Invasive Species the Means to Succeed? A Review of the Evidence." *Frontiers in Ecology and Evolution* (2020): 187. https://doi.org/10.3389/fevo.2020.00187.

Thomson, Jessica E., Sophie S. Hall, and Daniel S. Mills. "Evaluation of the relationship between cats and dogs living in the same home." *Journal of Veterinary Behavior* 27 (2018): 35–40.

Turcsán, Borbála, Lisa Wallis, Zsófia Virányi, Friederike Range, Corsin A. Müller, Ludwig Huber, Stefanie Riemer. "Personality traits in companion dogs—Results from the VIDOPET." *PLOS ONE* 13, no. 4 (2018): e0195448. https://doi.org/10.1371/journal.pone.0195448.

Turnbull, Jonathan. "Checkpoint dogs: Photovoicing canine companionship in the Chernobyl Exclusion Zone." *Anthropology Today* 36 (2020): 21–24. https://doi.org/10.1111/1467-8322.12620.

Úbeda, Yulán, Sara Ortin, Judy St. Leger, Miquel Llorente, and Javier Almunia. "Personality in Captive Killer Whales (*Orcinus orca*): A Rating Approach Based on the Five-Factor Model." *Journal of Comparative Psychology* 133 (2019): 253–61.

Van Lawick-Goodall, Hugo, and Jane van Lawick-Goodall. *Innocent Killers*. Boston: Houghton & Mifflin, 1971.

Vanak, Abi Tamim, Christopher R. Dickman, Eduardo A. Silva-Rodríguez, James R. A. Butler, and Euan G. Ritchie. "Top-dogs and underdogs: Competition between dogs and sympatric carnivores." In *Free-Ranging Dogs and Wildlife Conservation*, edited by Matthew E. Gompper, 69–87. Oxford: Oxford University Press, 2014.

Vanak, Abi Tamim, and Matthew E. Gompper. "Dogs *Canis familiaris* as carnivores: Their role and function in intraguild competition." *Mammalian Review* 39 (2009): 265–83.

———. "Interference competition at the landscape level: The effect of free-ranging dogs on a native mesocarnivore." *Journal of Applied Ecology* 47 (2010): 1225–32.

———. "Dietary niche separation between sympatric free-ranging domestic dogs and Indian foxes in central India." *Journal of Mammalogy* 90 (2009): 1058–65.

Vilà, Carles, and Jennifer A. Leonard. "Origin of Dog Breed Diversity." In *The Behavioural Biology of Dogs*, edited by Per Jensen, 38–58. Oxfordshire, UK: CAB International, 2007.

Vindas, Marco A., Marnix Gorissen, Erik Höglund, Gert Flik, Valentina Tronci, Børge Damsgård, Per-Ove Thörnqvist, Tom O. Nilsen, Svante Winberg, Øyvind Øverli, and Lars O. E. Ebbesson. "How do individuals cope with stress? Behavioural, physiological and neuronal differences between proactive and reactive coping styles in fish." *Journal of Experimental Biology* 220 (2017): 1524–32. https://doi.org/10.1242/jeb.153213.

Vonholdt, Bridgett M., Daniel. R. Stahler, Douglas W. Smith, Dent A. Earl, John. P. Pollinger, and Robert K. Wayne. "The genealogy and genetic viability of reintroduced Yellowstone grey wolves." *Molecular Ecology* 17, no. 1 (2008): 252–74.

Wallace-Wells, David. *The Uninhabitable Earth—Life after Warming.* New York: Tim Duggan Books, 2019.

Walsh, Bryan. *End Times: A Brief Guide to the End of the World.* New York: Hachette Books, 2019.

Wang, Xiaoming, and Richard H. Tedford. "Evolutionary History of Canids." In *The Behavioural Biology of Dogs*, edited by Per Jensen, 3–20. Oxfordshire, UK: CAB International, 2007.

Wayne, Robert K., and Stephen J. O'Brien. "Allozyme divergence within the Canidae." *Systematic Zoology* 36 (1987): 339–55.

Weisman, Alan. *The World without Us.* New York: St. Martin's Press, 2007.

Weiss, Alexander. "Personality Traits: A View from the Animal Kingdom." *Journal of Personality* 86 (2018): 12–22.

West-Eberhard, Mary Jane. "Phenotypic Plasticity." *Encyclopedia of Ecology* (2008): 2701–7.

Wheat, Christina Hansen, John L. Fitzpatrick, Björn Rogell, and Hans Temrin. "Behavioural correlations of the domestication syndrome are decoupled in modern dog breeds." *Nature Communications* 10 (2019): 2422.

Wiley, R. Haven, "Social Structure and Individual Ontogenies: Problems of Description, Mechanism, and Evolution." In *Perspectives in Ethology*, edited by P. P. G. Bateson and P. H. Klopfer, 105–33. Boston: Springer, 1981.

Wilkins, Adam S., Richard W. Wrangham, and W. Tecumseh Fitch. "The 'Domestication Syndrome' in Mammals: A Unified Explanation Based on Neural Crest Cell Behavior and Genetics." *Genetics* 197, no. 3 (2014): 795–808. https://doi.org/10.1534/genetics.114.165423.

Wolf, Max, and Franz. J. Weissing, "Animal personalities: consequences for ecology and evolution." *Trends Ecology and Evolution* 27 (2012): 452–61.

Woodyatt, Amy. "Is it a dog or is it a wolf? 18,000-year-old frozen puppy leaves scientists baffled." CNN, November 27, 2019. Accessed April 14, 2020. https://www.cnn.com/travel/article/frozen-puppy-intl-scli-scn/index.html.

Worboys, Michael, Julia-Marie Strange, and Neil Pemberton. *The Invention of the Modern Dog: Breed and Blood in Victorian Britain*. Baltimore: Johns Hopkins University Press, 2018.

Young, Julie K., Kirk A. Olson, Richard P. Reading, Sukh Amgalanbaatar, and Joel Berger. "Is Wildlife Going to the Dogs? Impacts of Feral and Free-roaming Dogs on Wildlife Populations." *BioScience* 61 (2011): 125–32. https://academic.oup.com/bioscience/article/61/2/125/242696.

PHOTOGRAPH CAPTIONS

(all photos by Marco Adda)

Chapter 1. A free-ranging Bali juvenile dog scavenging on daily offerings.

Chapter 2. A free-ranging adult female Bali dog provides alloparental care to a puppy dumped at the beach (Batubolong Village).

Chapter 3. Free-ranging Bali dogs form packs during home-range roaming (Batubolong Village).

Chapter 4. A free-ranging dog mum is occasionally tied in a courtyard to guarantee parental care to her puppies.

Chapter 5. Free-ranging Bali dogs mating. The mating male dog (brown dog at the center-right) protects his space, not allowing the other two males (white dogs) to approach the female dog (center-left).

Chapter 6. Free-ranging Bali dogs running and playing in Batubolong beach.

Chapter 7. Free-ranging Bali dog scavenging on daily offerings in a parking lot.

Chapter 8. A litter of free-ranging Bali dogs in a cave.

Chapter 9. Free-ranging Bali dog mum and her puppies during parental care and breastfeeding.

INDEX

Note: Page numbers in *italic type* refer to illustrations.

abuse, 32, *146*, 150
activity rhythms, 31, 50, 54–55
affective experience. *See* emotional skills
Afghan dogs, 57
African wild dogs (*Lycaon pictus*), 69, 93, 102
Akitas, 6, 49
Alaskan malamutes, 6, 52, 73
Allen's rule, 49
alloparental care, 76. *See also* parenting behaviors
American Pet Products Association (APPA), 170n18, 171n19
Animal Cognition (publication), 66
Animal Planet's Dog Breed Selector, 36
animal shelters: diet in, 60; dog population in, 27, 136, 169n15; killing of healthy animals in, 183n6; pandemic effects on, 182n5; posthuman changes to, 145, *146*
Antarctica, 26, 158
anti-predatory strategies, 65, 66
anxiety, 39, 85, 117, *146*, 150. *See also* psychological factors
APPA (American Pet Products Association), 170n18, 171n19
Arctic foxes, 58, 85
Arluke, Arnold, 171n24
Atema, Kate, 171n24
Australian cattle dogs, 6
Australian dingoes, 34–35, 72, 102, 135
AVMA, 182n4

baboons, 123, 182n23
barking, 88–89, 98, 101, 142, 144, 150. *See also* communication
basset hounds, 51
Bates, Amanda, 49
Beck, Alan, 94, 99
behavioral characteristics and skills: of Canidae family, 20–22; contact behaviors, 91; denning, 73–74; domestication and, 25; flexibility of, 108, 124, 125; human dependency of dogs, 13, 15, 28–30, *31*; parenting skills, 74–77, 153; posthuman survival and, 142, 148–49, 156–57; predatory behaviors, 4, 63–66, 87, 127, 142, 144, 174n6; problem solving, *43*, 110–15; scavenging strategies, 7, 27, 55, 63, 67, 103; sniffing, 71, 83, 129, 142; submission, 97; urine marking, 90–91, 101, 159, 178n11. *See also* ecological and evolutionary trajectories of

behavioral characteristics and skills (*continued*)
posthuman dogs; learning and skills development; socialization and social relationships
The Behavioural Biology of Dogs (Jensen), 29
Belarus, 80
Benson-Amram, Sarah, 111
Bergmann's rule, 48
Bernese mountain dogs, 37
Bhadra, Anindita, 67, 75, 76
bichon frise, 59
The Biology and Conservation of Wild Canids (Sillero-Zubiri), 167n1 (ch. 2)
body shape of dogs, 50–52. *See also* physical characteristics of dogs
body size of dogs: posthuman survival and, 3–4, 5, 6, 9–10, 37, 46, 47–50; reproduction and, 72; variation among, 22, 48. *See also* physical characteristics of dogs
Bonnani, Roberto, 75, 78, 97, 100
border collies, 6, 36, 65
boxers, 53, 133
Boydston, Erin, 104
brachycephaly, 53–54, 124, 130, 133, *146*. *See also* physical characteristics of dogs
Brazil, 180n37
breeding. *See* forced breeding; reproduction
Bremner-Harrison, Samantha, 118
Brisbin, Lehr, 78
Brooks, David, 153
Bryce, Caleb, 52
bulldogs, 6, 53, 70, 105, 133
bullmastiffs, 50
Burt, William, 98
Byrne, Richard, 115

Cafazzo, Simona, 75, 78, 97, 100
caloric needs, 67–68. *See also* diet
cancer, 133. *See also* disease
Canidae family, 17–22, 168n3. *See also* domestic dogs
Canids of the World (Castelló), 17, 34, 157
Canis genus, 19
Canis lupus familiaris classification, 19
Carnivora order, 19
Carnivores of the World (Hunter), 17
Carolina dogs, 78
Carter, Allisa, 114
Castelló, José, 17, 34
cats: cat-dog relationships, 5, 102, 103; classification of, 19; impacts on wildlife by, 180n37; posthuman survival and, 3, 4; prophylactic killing of, 140
Chaser (dog), 112–13
chasing behavior, 4, 65, 127, 142, 144, 174n6. *See also* behavioral characteristics and skills; predatory behaviors
Chernobyl, 4, 167n3
Chihuahuas, 3, 37
Chile, 170n18
climate and posthuman survival, 48–49, 56
climate change, 12, 49–50
climatic selection, 58
coats of dogs, 57–59. *See also* physical characteristics of dogs

cognitive ecology, *43*, 66–67, 108, 115–16. *See also* ecological and evolutionary trajectories of posthuman dogs
cognitive skills, 106–15, 128, 132. *See also* emotional skills; learning and skills development
"The Communal Organization of Solitary Animals" (Leyhausen), 83
communication: via ears, 55–56, 134; via eyes, 54, 88; posthuman survival and, *42*, 87–91; via tails, 56–57, 73, 134; through barking, 88–89, 98, 101, 142, 144, 150. *See also* socialization and social relationships
Conron, Wally, 133
contact behaviors, 91. *See also* behavioral characteristics and skills
convergent evolution, 173n5
Cooke, Robert, 49
cooperation, 5, 113–15. *See also* behavioral characteristics and skills; cognitive skills; socialization and social relationships
coping strategies, *44*, 120–21. *See also* emotional skills; psychological factors
core area, defined, 98. *See also* spatial use
corgis, 51
courtship, 71–72. *See also* reproduction
COVID-19 pandemic, 139, 182n5. *See also* disease
coyotes: behavioral characteristics and skills of, 21, 122; communication of, 88; diet of, 174n1; coyote-dog breeding, 5, 72, 135, 182n3; coyote-dog relationships, 102, 103, 104; personalities of, 85; reproductive cycle of, 7, 79; taxonomy of, 19. *See also* Canidae family
Creel, Scott, 68–69

dachshunds, 50
dalmatians, 57
Daniels, Thomas, 25, 63, 74, 79, 94, 100
Darwin's foxes, 21
demographics. *See* population size
denning, 73–74. *See also* behavioral characteristics and skills
dependency of dogs on humans, 28–30, *31*, 160. *See also* independence of dogs from humans
depression, *146*, 150. *See also* psychological factors
Derr, Mark, 5, 6, 35
Desert Dog (Kjelgaard), 175n8
desexing, 62, 69, 90–91, 137, 138, 178n11, 182n4. *See also* reproduction
diet: of coyotes, 174n1; of domestic dogs, 60–61, 174n1; of domestic *vs.* wild species, 25; of feral dogs, 61–62; of free-ranging dogs, 61–62, 63, 67; of homed dogs, 61, 62–63; posthuman survival and, 5, *41*, 50, 61–69, 124, 127; reproductive choices and, 76–77; scavenging strategies, 7, 27, 55, 63, 67, 103. *See also* predatory behaviors
dingoes, 34–35, 72, 102, 135
directional selection, *24*. *See also* natural selection
disease, 28, 80, 133, 139, 150, 169n15, 182n5
disfigurements, 133–34, *146*. *See also* maladaptive traits

disruptive selection, *24*. *See also* natural selection
dog breed, as concept, 6, 20, 35–39, 72, 133, 172n34. *See also* forced breeding; inbreeding; reproduction
dog fighting (sport), *146*, 150, 153
dog meat farming, *146*, 150, 161
Dogor (prehistoric dog), 23
dog populations. *See* population size
dog racing (sport), 50, 64, 65, *146*, 150
dogs and humans. *See* human-dog relationships
domestication, 22–24; defined, 33; physical characteristics and, 25, 52–53, 57, 58; scientific inquiry into, 162–63; sexual maturation and, 77; social characteristics and, 25–26, 152
Domestication: The Decline of Environmental Appreciation (Hemmer), 50
domestic cats. *See* cats
domestic dogs: classification of, 17–19; diet of, 174n1; impacts on wildlife by, 65–66, 179n37. *See also* Canidae family; feral dogs; free-ranging dogs; homed dogs; secondarily wild dogs, defined
Doomsday Preppers (TV series), 126
Drea, Christine, 114
Dunbar, Ian, 90

ears, 55–56, 134. *See also* physical characteristics of dogs
ecological and evolutionary trajectories of posthuman dogs, *41–44*, 115–25. *See also* cognitive skills; emotional skills; learning and skills development; physical characteristics of dogs
ecological niches of dogs, 28–30, 116, 157–59
Eigenbrod, Felix, 49
Elwood, Robert, 118
emotional intelligence, *43*, 116–18. *See also* intelligence
emotional skills, 106–15. *See also* behavioral characteristics and skills; cognitive skills; coping strategies; learning and skills development; play; psychological factors; stress responses
energy expenditure, 68–69, 77. *See also* diet
ethical issues: of forced breeding, 133, 137, 141, 150, 161; in a posthuman world, 125; of prophylactic killing, 139–40, 150, 183n6
euthanasia, as concept, 183n6. *See also* prophylactic killing
evolutionary tree of Canidae, 17, *18*, 19
eyes, 54, 88. *See also* physical characteristics of dogs

fatherhood. *See* paternal behaviors
Feddersen-Petersen, Dorit, 96, 97
feeding strategies. *See* diet
fennec foxes, 21
feral dogs: defined, 33–34; diet of, 61–62; on Navajo Nation, 74, 79, 100–101; population data on, 27, 169n15; posthuman survival skills of, 13, 31, 127, 140, 163; reproduction among, 61; scientific research on, 10–11, 158; social organization of, 94–95; stressors of, 120; territories of, 99. *See also* domestic dogs

The First Domestication (Pierotti and Fogg), 6
First-generation dogs, as group, 13, *14*
Fogg, Brandy, 6
food. *See* diet
forced breeding: ethical issues of, 133, 137, 141, 150, 161; hitchhiker traits from, 23–24; hybridization speculation on, 134–35; maladaptive traits from, 6, 70, 96, 132–34, *146*, 161; for the Superdog, 129–32. *See also* dog breed, as concept; natural selection; reproduction
Fox, Michael, 167n1 (ch. 2)
foxes: behavioral characteristics and skills of, 65; classification of, *18*; fox-dog relationships, 102; personalities of, 85; physical characteristics of, 21, 58; socialization of, 86–87
Francis, Robert, 33
free-ranging dogs: courtship of, 71; COVID-19 pandemic effects on, 183n5; defined, 32–33, 171n24; denning by, 74; diet of, 61–62, 63, 67; Fox on, 167n1 (ch. 2); learned skills of, 109–10; mortality of, 75, 79, 80–81, 177n39; parental behaviors of, 75–76; population data on, 28, 169n15, 170n18, 171n20; posthuman survival skills of, 127, 140; reproduction among, 61; scientific research on, 10–11, 158; social organization of, 92–93, 94, 95–96, 97; stressors of, 120; territories of, 99. *See also* domestic dogs
Free-Ranging Dogs and Wildlife Conservation (Gompper), 100
French bulldogs, 53, 105
functional morphology, *41*
fur, 57–59. *See also* physical characteristics of dogs

gains for dogs of a posthuman world, 145–53
Gardner, Howard, 180n1
gastric torsion, 133
general adaptation syndrome (GAS), 120
genetic diversity, 133, 134–36. *See also* forced breeding; inbreeding; reproduction
German shepherds, 36, 53, 133
German shorthaired pointers, 65
golden retreivers, 133
Gompper, Matthew, 100, 170n18
Gordon setter, 59
The Grass Library (Brooks), 153
Great Danes, 46, 133
The Great Dog and Cat Massacre (Kean), 140
greyhounds, 49, 50, 58, 64, 65
Griffin, Donald, 108
guilds, 102–3. *See also* socialization and social relationships

habituation, 177n3
Heid, Markham, 4–5, 7
Hemmer, Helmut, 50, 57
herding behavior, 65. *See also* behavioral characteristics and skills
Herzog, Hal, 170nn17–18
heterosis, 135, 139, 175n16
heuristics, 66–67
hip dysplasia, 133. *See also* maladaptive traits
Holekamp, Kay, 111

homed dogs: courtship of, 71; defined, 31–32; diet of, 61, 62–63; learned skills of, 109; neighborly relationships of, 102–5; population data on, 28, 169n15, 171n20; posthuman survival skills of, 127–29; scientific research on, 159; stressors of, 120, 150–51. *See also* domestic dogs; human-dog relationships
home range, defined, 98. *See also* spatial use
horses, 50, 122
hounds, 6
"How Dogs Would Fare Without Us" (Heid), 4
How the Dog Became a Dog (Derr), 5
human-dog relationships: improvements to, 141–42, 160–62; living arrangements and niches in, 28–30, *31*, 82; oxytocin and, 54, 117, 152, 173n15; paedomorphism and, 53–54; play and, 122–23; population spread and, 26–27. *See also* domestication; homed dogs; socialization and social relationships
hunting. *See* predatory behaviors
huskies, 6, 37, 49, 142
hybridization, 10, 72–73, 90, 134–36, 179n37. *See also* natural selection; reproduction
hyenas, 111, 114
hypothalamic-pituitary-adrenal (HPA) axis reactivity, 121

inbreeding, 72, 133, 175n15. *See also* dog breed, as concept; forced breeding; purebred dogs; reproduction
independence of dogs from humans, 7, 30, 60–61, 77, *146*, 151, 154, 163. *See also* dependency of dogs on humans
India, 67, 74, 75, 80, 81
intelligence, *43*, 180n1. *See also* emotional intelligence
International Union for Conservation of Nature, 180n37
interpersonal conflict, 115, 117, 128, 151–52. *See also* socialization and social relationships
Irish setters, 134–35
Italy, 75, 95–96, 172n2

jackals, *18*, 19, 72, 85, 102, 135, 182n3
Japan, 170n18
Jensen, Per, 23, 29

Kaminski, Juliane, 54
Kean, Hilda, 140
komondors, 57, 59
Koolhaas, Jaap, 121
K-strategy, 176n25

laboratory research experiments using dogs, 145, 147, 150, 159, 161. *See also* scientific research on dogs
Labradoodles, 133
landraces, 36–37
Later-generation dogs, as group, 13, *14*
Lazzaroni, Martina, 111
learning and skills development, *43*, 108–10, 132. *See also* behavioral characteristics and skills; ecological and evolutionary trajectories of posthuman dogs; emotional skills; problem solving skills

leptospirosis, 28. *See also* disease
Leyhausen, Paul, 83
life expectancies, 80. *See also* mortality
losses for dogs of a posthuman world, 145, *147*, 147–53

Macdonald, David, 20–21, 68–69, 167n1 (ch. 2)
Majumder, Sreejani Sen, 74
maladaptive traits, 6, 70, 96, 132–34, *146*, 161. *See also* forced breeding; physical characteristics of dogs
malamutes, 6, 52, 73
Marshall-Pescini, Sarah, 114
mastiffs, 6, 175n16
maternal behaviors, 74–77. *See also* behavioral characteristics and skills; reproduction
mating. *See* reproduction
Mech, L. David, 80
Mills, Gus, 68–69
Minnesota, 80
mixed-breed dogs, 38, 70, 134, 172n34. *See also* dog breed, as concept; reproduction
Morey, Darcy, 24, 29, 176n25
mortality: of free-ranging dogs, 75, 79, 80–81, 177n39; human-induced, 139–40, 150; life expectancy rates, 80–81; of lone coyotes, 122
multiple intelligences, as term, 180n1

natal philopatry, 101–2
National Geographic, 126
natural selection, 5–6, 13, *24*, 71, 138–39. *See also* forced breeding; hybridization; reproduction
Navajo Nation, 74, 79, 100–101
neutered dogs, 90–91, 178n11, 182n4. *See also* desexing
Newfoundlands, 6
New Zealand, 180n37
Norwegian elkhounds, 52

object manipulation, 111
opportunistic, as term, 7
ownership of dogs, as concept, 171n23. *See also* homed dogs
oxytocin, 54, 88, 117, 152, 173n15

Pacific Islands, 170n18
packs, 92–95. *See also* socialization and social relationships
paedomorphism, 53, 54
Pal, Sunil Kumar, 75
pandemics, 139, 182n5. *See also* disease
parenting behaviors, 74–77, 153. *See also* behavioral characteristics and skills; reproduction
parvovirus, 28. *See also* disease
paternal behaviors, 11, 75–76. *See also* behavioral characteristics and skills; reproduction
Paul, Manabi, 75, 76, 80, 81
peeing. *See* urine-marking
Pekingese, 46
persistence, 111–12
personalities of animals: breeds and, 36; posthuman survival and, 4, *44*, 118–19; research on, 181n13
pet, as term, 32
pet dogs. *See* homed dogs
phenotypic plasticity, 45–46, 130, 131
Philippines, 170n18

phylogenetic tree of Canidae, 17, *18*, 19
physical characteristics of dogs: body shape, 50–52; brachycephaly, 53–54, 124, 130, 133, *146*; of Canidae family, 20; domestication effects on, 25, 52–53, 57, 58; ears, 55–56, 134; eyes, 54, 88; fur, 57–59; maladaptive traits, 6, 70, 96, 132–34, *146*, 161; posthuman survival and, 46, 124, 129–32, 141–42; saggital crest, 25, 46, 52–53, 162; skulls, 51, 52–55, 124, 133; tails, 56–57, 73, 107, 134; teeth, 53, 89. *See also* body size of dogs; ecological and evolutionary trajectories of posthuman dogs
physical gains and losses for dogs in a posthuman world, 145–51
Pierotti, Raymond, 6
play, 86–87, 121–25, 128. *See also* emotional skills; socialization and social relationships
pointing, 65. *See also* predatory behaviors
Pongrácz, Péter, 76
poodles, 96
Poppy (dog), *130*
population size: of dogs, 26, 168n15, 169n16, 170n18, 171n19; of feral dogs, 27, 169n15; of free-ranging dogs, 28, 169n15, 170n18, 171n20; of homed dogs, 28, 169n15, 171n20; human-dog ratios, 27, 170n17; posthuman survival and, *41*, 148; of purebred dogs, 35–36, 172n34; of wolves, 26; zero growth scenario, 136–38
posthuman, as term, 12
posthuman survival of dogs, overview, 2–12, 143–55
predatory behaviors, 4, 63–66, 87, 127, 142, 144, 174n6. *See also* behavioral characteristics and skills; diet
preppers, 126–27
problem solving skills, *43*, 110–15. *See also* behavioral characteristics and skills; learning and skills development
Prodohl, Paulo, 118
prophylactic killing, 139–40, 150, 183n6
psychological factors: anxiety, 39, 85, 117, *146*, 150; coping strategies, *44*, 120–21; depression, *146*, 150; of homed dogs, 32; stressors, 120, *146*, 150–51; stress responses, *44*, 120–21. *See also* emotional skills
psychological gains and losses for dogs of posthuman world, 145–49, 152–53
pugs, 53, 89, 104, 133
puppies: parental behaviors toward, 74–77; socialization of, 84–87, 177n2
puppy mills, *146*, 150, 153, 161
Purcell, Brad, 35
purebred dogs: misidentification as, 172n32; population data on, 35–36, 172n34; posthuman loss of, 70, 135. *See also* dog breed, as concept; forced breeding; inbreeding; reproduction

rabies, 28, 80, 150, 169n15
Range, Friederike, 114
rat terriers, 5
red foxes, 21, 65, 85, 86–87
reproduction: denning, 73–74; diet and, 76–77; dog-coyote breeding, 5,

72, 135, 182n3; dog-dingo breeding, 72, 135; dog-jackal breeding, 72, 135, 182n3; dog-wolf breeding, 72, 90, 135, 182n3; for genetic diversity, 134–36; inbreeding, 72, 133, 175n15; posthuman survival and, *41*, 69–72, 77–81, 124, 141; sex ratios in, 79–80; Superdog breeding, 129–32. *See also* desexing; forced breeding; inbreeding; socialization and social relationships
reproductive cycles, 7, 10, 20, 22, 77–79
research laboratories using dogs, 145, 147, 150, 159, 161
research studies on dogs, 10–11, 157–60
Risch, Thomas, 78
Ritchie, Euan, 65
rope-pulling tasks, 13–14, 115
Rowan, Andrew, 170n18, 171n20
r-strategy, 176n25
running, 51–52

sagittal crest, 25, 46, 52–53, 162. *See also* skulls
Saint Bernards, 6, 50
salukis, 49
Santicchia, Francesca, 119
Sarkar, Rohan, 67
Sau, Shubhra, 67
Saudi Arabia, 27, 170n18
scavenging strategies, 7, 27, 55, 63, 67, 103. *See also* diet
scent marking, 90–91, 101, 178n11. *See also* behavioral characteristics and skills
scents, 66, 72, 90, 178n11
Schenkel, Rudolf, 97
scientific research on dogs, 10–11, 157–60. *See also* laboratory research experiments using dogs
secondarily wild dogs, defined, 34–35. *See also* domestic dogs
selective breeding. *See* forced breeding
self-determination. *See* independence of dogs from humans
sex ratios, 79–80
sexual abuse, 32, *146*, 150
sexual maturation, 25, 77, 176n26. *See also* reproduction
Seyle, Hans, 120
shared goal tests, 113–14, 115
shar-peis, 59, 133
sheepdogs, 58, 59
shelters. *See* animal shelters
Shiba Inu, 73
shih tzu, 53, 59
Sillero-Zubiri, Claudio, 20–21, 167n1 (ch. 2)
size. *See* body size of dogs
skin diseases, 133. *See also* disease
skulls, 51, 52–55, 124, 133. *See also* physical characteristics of dogs
Smith, Bradley, 35
sniffing behavior, 71, 83, 129, 142. *See also* behavioral characteristics and skills
Sober, Elliott, 23
social cognition, *43*, 128
socialization and social relationships: cooperation, 5, 113–15; defined, 83; domestic effects on, 25–26; *vs.* habituation, 177n3; human control of, 61, 83, 88, 113, 128, 178n19; interpersonal conflict in, 115, 117,

socialization and social relationships (*continued*)
128, 151–52; neighborly relationships, 102–5; play, 86–87, 121–25; posthuman survival and, *42*, 91–97, 128–29, 145–49, 151–52; of puppies, 84–87, 177n2. *See also* behavioral characteristics and skills; communication; human-dog relationships; reproduction
Societies of Wolves and Free-ranging Dogs (Spotte), 11, 56, 99
South Africa, 69
South Asia, 170n18
spatial use, *42*, 98–102, 178n19
spayed dogs, 90–91. *See also* desexing
speculative biology, 8, 10, 157
spitz, 73
Spotte, Stephen, 11, 56, 80, 99
squirrels: diseases of, 119; hunting of, 108, 142, 174n6
stabilizing selection, *24*, 57, 79. *See also* natural selection
Staffordshire terriers, 50
stalking behavior, 4, 87. *See also* behavioral characteristics and skills; predatory behaviors
Stearns, Stephen, 44
stray dogs. *See* feral dogs; free-ranging dogs
stressors, 120, *146*, 150–51. *See also* psychological factors
stress responses, *44*, 120–21. *See also* emotional skills
submissive behavior, 97. *See also* behavioral characteristics and skills
Superdog, development of, 129–32
surgical disfigurements, 133–34, *146*
Sweden, 170n18
sympatric relationships, 102–4. *See also* socialization and social relationships
Sztruhala, Sára, 76

tails, 56–57, 73, 107, 134. *See also* physical characteristics of dogs
tapetum lucidum, 54
teeth, 53, 89. *See also* physical characteristics of dogs
territory, defined, 98. *See also* spatial use
The Theory of Island Biogeography (Wilson and MacArthur), 176n25
The Thinking Ape (Byrne), 115
Tibetan mastiffs, 175n16
Transition dogs, as group, 13, *14*
Turnbull, Jonathan, 167n3

United Kingdom, 170n18
United States, 27, 170n18, 171n19
urine-marking, 90–91, 101, 159, 178n11. *See also* behavioral characteristics and skills

Vanak, Abi, 103
village dogs, defined, 33. *See also* free-ranging dogs
vocalization. *See* barking; communication

Weimaraners, 57
Weisman, Alan, 2–3, 5
whippets, 50
The Wild Canids (Fox), 167n1 (ch. 2)
wild cats. *See* cats

wild dogs. *See* African wild dogs (*Lycaon pictus*)
wildlife impacts of dogs, 65–66, 179n37. See also *names of specific species*
Williams, Terrie, 52
wolverines, 83, 158
wolves: behavioral characteristics and skills of, 21; communication of, 88, 89; courtship of, 71; diet of, 62, 68; learned skills of, 109; mortality rates of, 81; reintroduction of, 103–4; reproduction of, 77–78, 175nn15–16, 182n3; reproductive cycle of, 7; sex ratios in, 79–80; social organization of, 92, 93; taxonomy of, 19; territories of, 101; wolf-dog breeding, 72, 90, 135, 182n3; wolf-dog relationships, 102, 103, 114. *See also* Canidae family
World Health Organization (WHO), 169n15
The World without Us (Weisman), 2–3

zero-growth scenario, 136–38